D1454497

THE TRAWLERMEN

by David Butcher

Tops'l Books
9 Queen Victoria Street,
Reading, England

Cover picture: Shooting the trawl. The belly and batings being got overboard on a deep sea trawler.
Below: *Boy Sam LT1014,* a Lowestoft sailing trawler of around 1910. The pronounced forward rake of the mizzen mast was a feature of the Lowestoft smacks.

Published 1980 by

Tops'l Books,
9, Queen Victoria Street,
Reading RG1 1SY
Berkshire, England

Printed in Great Britain by
Lonsdale Universal Printing Ltd

British Library CIP Data

Butcher, David Robert
The trawlermen.
1. Fisheries – England – East Anglia –
History – 20th century
2. Trawls and trawling – England –
East Anglia – History – 20th century
338.3'72'709426   SH258.E1

ISBN   0 906397 06 5   (hardback)
ISBN   0 906398 05 7   (paperback)

Also from Tops'l Books
Sailing Fishermen In Old Photographs
by Colin Elliott
Steam Fishermen In Old Photographs
by Colin Elliott
The Driftermen
by David Butcher
Provident and the Story
of the Brixham Smacks
by John Corin

Also by David Butcher

Waveney Valley
(East Anglian Magazine Ltd)

Opposite:
Smacks leaving Lowestoft harbour for the North Sea grounds around 1910.
Ahead of them, one of the early steamers with 'Woodbine' funnel set in
front of the wheelhouse. Astern is *Snowdrop LT543* and ahead of her
Search LT5.

For the Trawlermen of Great Britain, then and now.

Illustrations

The photographs on pages 35, 62, 70 (lower), 88 (lower), 107 and 114 are reproduced by permission of The Times Picture Library. The portraits on page seven were specially taken by Rick Turrell. Mr Charles Crisp kindly loaned the photographs of his father and brother on page 42. All other photographs in the book are the copyright of Ford Jenkins, Lowestoft.

Line illustrations by
Syd Brown

The text of this book contains many specialised local, technical and nautical words which may not be familiar to the general reader. To avoid peppering the pages with footnotes and explanations a full glossary is included at the back.

Contents

Opposite: the Grimsby trawler *Aston Villa GY42*, one of a number of the Consolidated company's fleet named after famous football teams, a steam trawler of 192 tons, 750 h.p., built at Beverley in 1946. Astern of her *Northern Gift GY704,* 194 tons and 1400 h.p., also built at Beverley for Northern Trawlers Ltd.
Page Eight: knee deep in fish as the cod end is hoisted inboard and emptied at the end of a haul.

Preface

Like my previous book, *The Driftermen,* this tale is based not on documents and written records, but on the first hand living memories of the men involved. Thus *The Driftermen* set down an authentic account of the now vanished herring industry, once one of the staple industries of Britain. *The Trawlermen* attempts to do the same for the other important branch of the fishing industry. Theirs was a different method, a different quarry and a different way of life. They shared in common with the drifter folk the perils of the sea, endless hard graft and lousy conditions for scant reward, and a salty tolerance of life as it came.

Unlike the herring fishing, trawling is not extinct. It is, however, a pale shadow of its former self and its future is uncertain. It is therefore important to set down now the experiences of the men who knew it in its heyday, in the first half of this century when British trawlers dragged and scoured the sea bed, round our coasts and anything up to a thousand miles away beyond the Arctic Circle.

Theirs is the authentic voice. It is heard in the book in their own expressive and vigorous language and, apart from providing a short introduction to set the scene for each chapter, I have tried to 'butt in' as little as possible. The rich East Anglian accents I have also tried to leave intact, consistent with the reader from 'foreign parts' being able to follow the sense.

I would like to take this opportunity of thanking all my subjects for the interest they have shown in the whole project and for the hospitality they and their wives have extended me whenever I have called. I value these meetings and like to think that the quality of the tape-recordings reflects the rapport that exists between interviewer and subject. One of them once said to me, 'You're like a bloody detective, with that notebook o' yours!' If any detective carries out investigatory work half as enjoyable as the kind I'm involved in, he is a fortunate man.

Besides the men themselves, I must also express my gratitude to a number of other people. First of all, there is the book's illustrator, Syd Brown, whose drawings complement the text so beautifully. Next comes Roy Harden Jones of the Fisheries Laboratory at Pakefield, who has done much to instruct me in the more scientific aspects of fishing. George Ewart Evans, a pioneer of oral history writing, continues to take an interest in my activities and so does my publisher, Colin Elliott. I am indebted to David Wright and his colleagues in the reference section of the Lowestoft Library for help in tracking down details of various vessels, and to members of the Lowestoft Port Research Society for access to their archives. Thanks are also due to Peter Jenkins for allowing me to search for photographs in his wonderful family collection. Peter Leighton, Secretary of the Lowestoft Fishing Vessel Owners Association, provided very helpful figures on the state of the trawling fleet in the port over a 30 year period, and Captain S. T. Smith, Editor of Olsen's Almanac, supplied from his prodigious knowledge a number of details that I wasn't able to get from local sources. To all of these people, and to many more not named, a sincere thank you.

The Men Who Told The Tale

ERNIE ARMES – born 1902 into a fishing family. Worked for 50 years on the Lowestoft market and has an encyclopaedic knowledge of the fishing industry. Very interested in local history.

HARRY COLBY – born 1902. Left his native Pakefield as a young man and spent 30 years trawling out of Fleetwood. Very active in retirement as treasurer of the Lowestoft Retired Fishermen's Club.

HERBERT DOY – born 1900. Went smacking as a young lad and later had several years in drifter-trawlers during the 1920s and 30s. Also worked on board Ministry research vessels.

ARTHUR EVANS – born 1901, the son of a leading trawl-fish merchant and boat-owner. Still takes an active interest in his business and also in the Lowestoft Rotary Club.

TED FENN – born 1898, the son of a horseman. Volunteered for naval service in World War I and served in armed smacks.

BILL JARVIS – born 1906. A Gorleston man, who began fishing in his teens. An active member of the Lowestoft Retired Fisherman's Club and the Naval Patrol Service Association.

NED MULLENDER – born 1896 into a Pakefield fishing family. First went to sea in 1910 and, during the course of a long career, had experience of drift-netting, long-lining, seining and trawling.

JACK ROSE – born 1926, the son of a longshoreman. Went fishing as a youngster in motor smacks. A dedicated and far-sighted collector of Lowestoft history.

GEORGE STOCK – born 1903 into a local fishing family. Always a trawlerman first and foremost, he moved to Fleetwood in the 1930s and was a successful skipper there for many years.

HORACE THROWER – born 1904. Began fishing at the age of 13 and spent all his working life at sea. For most of the time he was a stoker on board steam drifters and drifter-trawlers.

CHAPTER ONE

Draw Shallow, Draw Deep
The Development of Trawling

'Sailing over the Dogger Bank –
Oh, wasn't it a treat!
The wind a-blowin' about east-north-east
So we had to give her sheet.
You ought to have seen us relish
The wind a-blowin' free,
A passage from the Dogger Bank
To Great Grimsby.'
(Traditional fisherman's chorus)

From the time of ancient Egypt onwards man has been dragging a variety of net bags along the bottom of rivers and oceans, in the hunt for fish which live in the depths. As with most of Man's most fundamental inventions little was recorded until the last hundred years. Bygone ages saw nothing remarkable in the way men earned their daily bread, at least nothing that merited being written down. It is only now, after some 200 years of industrialisation, that we feel cut off from our roots, from that organic part of ourselves which sees the importance of growing and gathering food. And with this feeling of severance, of loss, comes the desire to record what we can of the old ways.

With trawling the basic principle remained unaltered over centuries, but improvements in gear and method gradually brought greater efficiency, greater catching power.

The first innovation was that of keeping the mouth of the bag-net open with a wooden beam, to which the upper part of the net was nailed. The lower portion was weighted with stones and allowed to drag along the bottom, while the beam itself was fixed to two stout, iron frames (rather like gratings) at either end, which held it in position and prevented it from either rising or sinking. This 'wondyrchoun', as it was known (literally, a wondrous device), is first heard about in the reign of Edward III, and we are also told its exact dimensions. The figures are important because they describe in detail the first beam trawl to be mentioned in English history. Here they are: a beam of 10 feet in length and a net 18 feet long, tapering down from a 10 foot width, where it was attached to the beam and with meshes about three inches square.

It is significant perhaps that such a revolutionary contrivance should be mentioned only because of the trouble it caused! Edward was petitioned in 1376 about the decline of the fisheries in the Thames Estuary caused by the 'wondyrchoun' which, it was claimed, had been in use for about seven years. We are told that it was made in the fashion of a large oyster dredge, or 'scrope net' as it was usually called. The complaints have a modern ring about them. The new trawl was accused of catching immature fish and of destroying the breeding grounds and spawn of various species. Throughout trawling history similar protests have been raised as new methods have appeared. They are heard today about powerful continental motor vessels hunting the North Sea for soles and plaice. They have all too often proved valid.

Just how important the wondyrchoun became in the fishing industry of mediaeval England is hard to say because the great majority of demersal fish (i.e. those that live on or near the bottom) were caught by baited hooks tied to handlines and longlines. From about the end of the 13th century to the middle of the 17th there was a great summer voyage to Iceland that involved so many of the East Coast ports (especially in Suffolk and Essex) sending out boats for a five month period and hopefully receiving them back with good quantities of salted cod and ling. During the winter these boats would first of all refit and then fish nearer home in the North Sea, taking whiting and haddock on their lines, as well as cod – catches that were again salted and dried into what became loosely known as 'stockfish'.

Although the wondyrchoun, or early beam trawl, didn't become the tool of the bigger fishing boats for at least another 200 years, the smaller inshore craft obviously persevered with it because there are a number of references to its use being restricted in British coastal waters. For instance, the Suffolk port of Orford banned it from the haven in the year 1491, referring to it as an 'engine', a good mediaeval word for any mechanical device. Round about the same time (1499) the Flemish authorities were also frowning upon it in their own country, so obviously the trawling technique had spread around the North Sea edges. We hear more of it in England during the reigns of Henry VIII and Elizabeth I, and by the 17th century it was obviously widespread, for various commissions and acts of Parliament from the time of James I right through till Queen Anne sought to regulate its use and to control the size of the meshes. One official body, led by the Chief Justice, Sir Edward Coke, in 1631, declared the size of the trawl net at that time to be 24 feet wide on the beam and with a length of 48 feet. This is a lot bigger than the old wondyrchoun and obviously required a larger boat to tow it along.

Much of the activity at this time centred on the Essex town of Barking, which has a good claim to be considered the cradle of the British trawling industry. Brixham men contest that claim, but from the evidence of available records I am inclined to favour the Thames-side community. Leaving any regional rivalries aside, what is important is that a developing industry was beginning to take shape, with its catches becoming more and more necessary as a supplement to the traditional deep-sea fishery based on line and hook. This became especially true after the decline of the Iceland voyages towards the end of the 17th century, because this summer fishing had been a major source of food for something like 350 years.

The exploitation of Newfoundland cod during the 18th century helped to fill the gap, but it had a comparatively short life compared with the Iceland fishing and the English boats fell more and more to reaping the waters nearer home. True, the handline and longline remained important, but trawling was a method that became increasingly adopted for the banks of the southern North Sea and the English Channel. Barking reigned supreme in the former area, Brixham in the latter, particularly after about the middle of the 18th century. The Devon men tended to work a bigger beam trawl than their Essex counterparts, perhaps because of the greater depth of water in the areas they fished.

During the period of the Napoleonic Wars the country's developing trawling industry, based on Barking and Brixham, had to restrict its activities partly because of the depredations of French privateers and partly because many fishermen went into the Navy,

Above: hauling the trawl aboard the *Eta LT57*, one of seven small boats built in Richards yard at Lowestoft in the 1930's for a company named LT1934 Ltd.
Below: the steam trawler *Thomas Lawrie LO318* built at Chambers yard on Oulton Broad to Admiralty specifications just after the First World War. She is similar to the 'Castle' boats from Swansea described in chapter eight.

either as volunteers or pressed men. However, the wars did serve to keep the price of fish high because of the general shortage, and so it remained a profitable enterprise. When peace finally came in 1815 and the fishing fleets expanded again, the price of trawl fish dropped considerably. This had a particularly severe effect on Brixham, which was a long way from any major centre of population and had always found the transportation of its catches a big problem. Barking, being so much nearer to London, wasn't nearly so inconvenienced.

It was this drop in prices, together with the difficulty of transport, that seems to have persuaded many of the Brixham men to uproot themselves and move round to the Kentish ports of Ramsgate and Dover, settling there permanently. The shift wasn't a completely new venture because from about the middle of the 18th century boats from Devon had fished off Dungeness, particularly for the soles and turbot to be had there, because these two species were firm favourites in the rapidly growing habit of eating flatfish. A culinary trend, it should be added, that demanded the use of fresh fish, not ones that had been salted down. And so this combination of gastronomic fashion and post-war depression helped cause an important permanent migration of experienced fishermen from the Channel coast of Devon round to the east side.

So the stage was set for the age of large-scale trawling, with the North Sea as the premier fishing ground. For centuries the Dogger Bank had seen Dutchmen, Germans, Belgians, Frenchmen and Danes fishing for herring with drift nets and for demersal species with handline and longline. Now it was to witness an explosion in the use of the beam trawl, which drew large catches of fish from what were almost virgin areas — the mud and sand gulleyways and holes that intersperse the bumps and rises of the great bank. First to be discovered, in 1837, was the Little Silver Pit, about 30 miles east of Spurn Head, so much favoured by soles. It was a Brixham man who chanced upon it and his discovery was soon being exploited by large numbers of boats from the established centres of trawling — Brixham, Ramsgate and Barking.

Soon the whole of the western side of the Dogger Bank was opened up as the sailing smacks groped further in their exploration of the bottom. The Skate Hole, the Great Silver Pit, the Hospital Ground, the Botney Gut and many more were the names given to spots where fish was found in great abundance: sole, turbot, brill, plaice, cod, whiting and haddock for the most part, and all of it eminently marketable.

There was only one problem with this great harvest of the sea and that was the business of finding somewhere to sell it while still fresh.

In an attempt to reach a market more quickly some of the Devon families who had earlier settled in Ramsgate moved northwards, and there were some Kentish folk that went with them, to Hull, a port of centuries' standing. However, the port authorities gave the new arrivals very little encouragement in their enterprise. The fish quay had very little berthing space; the rail head was over a mile and a half away; and there were inadequate facilities for the housing of stores and the servicing of boats. Then came the great breakthrough. In 1848 the Manchester, Sheffield and Lincolnshire Railway began to develop Grimsby as a fishing port. The dock facilities and hiring rates offered were attractive, and there was the added advantage of the fishing boats not having to negotiate several miles of the Humber Estuary in order to dock. Moreover, the rail network connected the new port with the rapidly growing industrial population centre of the North and Midlands, and also with London. From a total of around 3000 in 1815, Grimsby's population rose to just over 11,000 in 1861. The number of fishing boats at this time was 315.

About the same time another important development in trawling took place. This was the organisation of fishing on a large scale, which became known as 'fleeting'. The technique was invented and perfected by Samuel Hewett of Barking, who owned a large number of fishing boats and who worked out a simple and efficient method of operation to compensate for the ever-increasing distances between the North Sea fishing grounds and his base in Essex. What he did was to keep his whole fleet at sea for anything up to a couple of months, fishing the productive Dogger Bank continuously. Catches were packed into large wooden cases known as trunks and then rowed across in dinghies to a fast-sailing cutter, which would load up and run the fish from the whole fleet to Billingsgate. Catches were sorted into two categories, prime and offal, the former consisting of soles, turbot, brill, and halibut, the latter of plaice, cod, haddock, whiting and all the rest. Ice was carried on board the cutters and packed between the trunks of fish to help keep the contents fresh. It was in fact Samuel Hewett who pioneered the use of ice in the fishing industry, his main source of supply being the frozen Essex marshes in the winter months.

The efficiency of 'fleeting', as opposed to having each trawler spend time sailing its own catch to port, soon became apparent and other operators began to employ it. By the 1860s and 70s it was almost universal, and efficiency improved further with the introduction of faster steam-powered fish-carriers. And not only did the large boatowners adopt the method; many of the smaller men did as well, attaching themselves and their vessels to the large fleets and working in with them. Yet if fleeting made economic sense, the toll it exacted in human terms was a heavy one. Working conditions on the boats (which were anything between 40 and 80 tons net) were absolutely appalling and gave rise to the expression 'sentenced to the Dogger'.

Among other things, it led to the founding of The Mission To Deep Sea Fishermen by E. J. Mather Esq. in 1881, which became one of the great secular ministries of the 19th century. After a century of service it is still active.

Other East Coast ports began to feel the benefit of the opening-up of the Dogger Bank. One of them was Great Yarmouth, which had had only the most passing acquaintanceship with trawling until Samuel Hewett moved his whole fleet up to Gorleston from Barking in 1854 so as to be nearer the main North Sea grounds. Hitherto, herring fishing had been Yarmouth's life-blood, but the introduction of Hewett's great Short Blue Fleet to the town on the opposite side of the Yare estuary sparked off a boom in trawling as well. From a mere handful of sailing trawlers in the 1840s, the fleet had arisen to over 400 by 1880, with at least three other big fleets besides Hewett's: The North Sea Trawling Co. Ltd., The Great Yarmouth Steam Carrying Co, Ltd., and Morgan & Co. Among the substantial private owners were E. A. Durrant, with around 35 boats, and Samuel Smith, with about 20. Fair holdings to be sure, but ones which tend to dwindle beside the sheer numbers of the Hewett fleet: 82 vessels in the year 1882, all of them between 35 and 65 tons net, and each crewed by six or seven men.

The Great Eastern Railway had arrived in Yarmouth in 1844, but although most of her trawlers operated as fleets out in the North Sea, the bulk of the catches went straight to Billingsgate on the cutters and never saw the home port. This wasn't true of the neighbouring town of Lowestoft which, after the railway arrived in 1847, began to expand quite rapidly, first of all as a herring port and then as a trawling base. In 1863 there were 176 herring luggers in Lowestoft and only eight sailing trawlers; by the end of the 19th century there were around 400 herring boats and 300 trawlers!

For some reason, the Suffolk port never adopted fleeting as a mode of operating, on such a large scale as other ports, perhaps because its trawlers remained in the hands of small syndicates and private individuals. One contributory factor in this pattern of boat ownership was no doubt the number of Ramsgate men, owner-skippers many of them, who moved up to Lowestoft in the middle of the 19th century to exploit the local grounds, which were less spectacular in their offerings than the Dogger Bank, but which yielded a decent living nevertheless. From the Gabbards and Galloper down south, up to the Leman and Ower in the north, and out eastwards to the Brown Ridges, there was plenty of fish to be had − and the competition for it wasn't so stiff and highly-organised. And so the Lowestoft pattern of trawling emerged: catches taken within reasonable sailing distance of the port, landed there, and then dispatched by rail to London, the Midlands and the North.

The other great development of the 19th century was of course the change from sail to steam. Converted paddle tugs were employed from 1877 but had too many limitations. The screw prop steam trawler soon emerged as a highly specialised form of vessel.

One of the first built was the *Zodiac GY825,* which had its keel laid down in a Hull shipyard in 1881 and was destined for a firm over the water in Grimsby. Her launching, in the December of 1881, ushered in a new era because she (and her sister boat, the *Aries GY832*) showed that their catching power was about three times that of a sailing trawler. Even at a cost of around £3500, these early steamers soon caught on because the investment was one that paid off handsomely. And it wasn't long before the wooden hulls gave way to iron ones, and the iron ones to steel. In about 15 years, sailing trawlers in Hull and Grimsby were a thing of the past.

This remarkable and rapid changeover wasn't solely due to the steam engine, however. The arrival of steam in the fishing fleets coincided with the invention of the otter trawl, the most important innovation in fishing gear since the 'wondyrchoun' of 500 years before. The first specific mention of an otter trawl was made in 1865 by J. Wilcock in his book, 'The Sea Fisherman Or Sea Pilotage', and while it had no doubt been going for some time before that, it is the last quarter of the 19th century that saw its general adoption. What it did was to spread the mouth of the trawl net in a different fashion. The beam trawl had the top of the net fixed to a spar of ash or beech anything up to 60 feet in length, thereby keeping it open, and two bridles (one at each end) then joined the beam to the towing warp. The otter trawl had two stout, iron-shod, wooden 'doors' at either end of the net, each on its own separate warp. As the boat was towing along, these doors floated upwards and outwards, thus opening the mouth of the net. It was a method that needed a good deal of power to make it work properly and thus it was ideally suited to the steamboats. Sailing vessels could work otter trawls, but only small ones, and this obviously meant small catches. By about 1890 most of the steamers had given up their beam trawls and gone over to otter gear. The age of industrial fishing had arrived!

Most of the new vessels were concentrated on the Humber, at Hull and Grimsby, but there were other sizeable trawling ports as well as North Shields, Milford Haven, Fleetwood and Aberdeen. North Shields had always been a traditional fishing centre, but Fleetwood didn't emerge until the mid 19th century, when its boats started to find a good market in the manufacturing towns of industrial Lancashire. With the rise of steam power, it found itself ideally placed close to a coalfield and with the whole of the west coast waters open to exploitation. Milford Haven was in much the same position, and once the Great Western Railway had built and opened a large fish dock there in the late 1880s it too was ripe for expansion. Aberdeen was some distance from any source of coal, but it developed as a considerable trawling port, the only one of any size in Scotland. Why that is so is debatable, but it may have had something to do with the fact that the other ports of north-east Scotland (Fraserburgh, Peterhead, Banff, Buckie etc.) were so committed to herring fishing. In fact, trawling was very unpopular with the Scottish driftermen because they claimed it destroyed the herring grounds.

By 1909 there were around 1340 steam trawlers in England and Wales (960 of them at Hull and Grimsby) and a further 200 or so at Aberdeen. The capital investment in such a programme of shipbuilding was obviously a large one, and one that the owners would want to see a return on. To begin with, the new steamboats from Hull and Grimsby continued to work on a fleet basis as their sailing predecessors had done, but by the beginning of the 20th century the Grimsby owners had largely abandoned this method of organising things and were now sending their boats further and further afield. Icelandic grounds were trawled for cod and there was a fishery established too in Russian waters, the Barents Sea proving a good hunting ground for any trawler prepared to make the trip. Some of the Hull fleets continued to work the Dogger Bank up to the time of the First World War, a good example being the Gamecock Fleet, which in 1912 consisted of 42 steam trawlers, six steam fish-carriers and a hospital ship. Two fleets continued to work the Dogger Bank till as late as 1936.

The years between 1900 and 1914 were the absolute peak of the steam trawling era, and it happened to coincide with the rise of what was to become recognised as the British national meal − fish and chips. There were 25,000 fish and chip shops in the country just before the Great War. Whether their presence helped to cause the steam trawling boom, or whether the vast catches landed by the boats fostered the frying trade is uncertain. What is certain is that the fish-friers weren't too bothered about the quality of the cod, haddock and plaice they handled because their batter and fat covered a multitude of sins in the cooking. And a lot of the white fish landed by the steam trawlers was inferior stuff, suffering first of all from rough treatment on the bed of the sea, being dragged along in the trawl, and later from the long run home from distant grounds, packed in ice down in the hold.

In the midst of all this activity, the steam trawler had very little effect on the East Anglian ports of Yarmouth and Lowestoft. Here sailing smacks continued to be economically viable, surviving in the latter place to serve a small quality market right down to 1939.

One Lowestoft firm, The Lowestoft Steam Trawling Co., did make an attempt to work steam boats from the port. In 1887 three vessels, the *Greencastle LT101,* the *Dolphin LT100* and the *Bonito LT106,* were purchased from Scotland and tried out with their beam trawls on the local grounds. They weren't a success, however, and were sold the following year. Trawling in Yarmouth suffered the final body blow in 1901, when the Hewett Short Blue Fleet ceased operations, unable to compete any longer with the steam trawler. It was the culmination of a downward trend of a good decade's length. The Yarmouth fleets just couldn't keep pace with the Hull and Grimsby steamers. They were further from the key fishing grounds; further from the essential supplies of steam coal and the owners weren't prepared to change their mode of operation. From around 600 sailing trawlers in 1887, Yarmouth's total had dropped to about 80 in 1903 — and most of these had gone by the outbreak of war in 1914. Lowestoft, on the other hand, which had a local-based trawling industry, supplying local markets as well as the big industrial centres, had about 300 sailing trawlers in 1887 and still 250 in 1914.

When the boats went back to fishing in 1919 and 1920 after the four-year stint of patrolling or minesweeping duties for the Admiralty, things in the fishing industry soon began to alter considerably, and the changes were mainly for the worse. This wasn't generally realised at first. After all, plenty of fish were being landed before the outbreak of war and there was no reason to think that things would be different once the conflict was over. True, trawling yields had gone down after the peak year of 1907, but not so much as to give cause for real alarm, and 1913 had been the absolute acme of achievement for the herring drifters. Now that things were starting up again, with the North Sea having had greatly reduced fishing for a four-year period, the general mood in the fish trade was one of optimism. But how quickly the dream was shattered. The collapse of many European currencies, and the loss thereby of traditional markets, crippled the herring trade; the over-intensive trawling from Hull and Grimsby cleaned out the North Sea — especially that great marine cornucopia, the Dogger Bank.

Against this background of falling catches, the economic state of Britain generally was very depressed and thus the 1920s and 30s saw the fishing industry in steady decline. Some attempt was made to compensate for the fall in North Sea catches by pushing trawlers up to the Faeroes, Iceland, Greenland, Bear Island and even Spitzbergen in search of the cod that the British housewife was convinced she must have. In fact, the one part of the fish trade that seems to have remained reasonably stable during this difficult time was the frying side. A crucial factor here (indeed, it may well have been the most important development in fish-processing this century) was the popularisation of filleting by William Eunson of Aberdeen. A native of Fair Isle originally, he established a business in the 'Granite City' and round about the end of the First World War pioneered the cutting of white fish into single fillets and block fillets.

The technique was just what the fried fish trade needed. The standard of the commodity which changed hands across the counter had always been variable, and now the friers could get away with even more. So too could the wet fish shops. Inferior fish could be made presentable as single fillets, while small stuff (hard to sell in a whole-fish trade) could look quite respectable as block fillets. It's hard to say how much damage the filleting technique did to our fishing grounds and stocks in the 1920s and 30s (and indeed after the Second World War) as the steam trawlers ploughed backwards and forwards, scooping up everything. Even if you finished up by being unable to sell what you'd caught, then the remedy was quite simple — either dump it, or sell it for fish-meal, and go out and get some more. Perhaps you'd be luckier next time and find a buyer.

Just as things were getting really bad, World War Two broke out and most of the fishing boats were again chartered by the government for patrolling or minesweeping. When they came back to fishing in 1946 the steam era was drawing to an end, and it wasn't long before the diesel trawlers which first appeared in the mid 30's, began to take over. The post-war period also saw the steady decline of the North Sea herring, until its fishery finally collapsed in the late 1960s. This left Yarmouth high and dry after a thousand years as a herring port, but the town was partly able to fill the gap by diversifying into new North Sea industries — especially the servicing and supply of the gas and oil rigs. Many of the boats seen in port now are working offshore in the new North Sea bonanza.

The culmination in 1976 of three long drawn out 'Cod Wars' with Iceland resulted in British fishermen being kicked out of waters they regarded as traditionally theirs and this has resulted in a good deal of hardship in Hull, Grimsby and Fleetwood. Many boats have been laid up (even some of the newer freezer stern-trawlers), others sold of for conversion to mid-water fishing, and it remains to be seen whether the fishing companies will be prepared to undertake the investment involved in seeking out new grounds in the Atlantic and trying to market hitherto unheard-of species.

At first, one port seemed to have benefitted from the misfortune of the other three, and this was Lowestoft. When Iceland closed, she found herself in the happy position of never having been a cod port and thus her mid-water boats were able to keep working on their post-war grounds off Denmark and Norway. In the autumn of 1979 there were around 70 of them fishing (including a number of Iceland boats brought down from Fleetwood and Grimsby) and these, when added to the inshore fleet of some 40 craft, made Lowestoft the country's second fishing port. No one in the town, though, saw this as the beginning of a new golden age in the fishing industry. The ever-rising cost of fuel and competition in home waters from our European partners make a boat's profitability uncertain. So does the sale of foreign fish on the local market. It is factors like these that have persuaded many fishermen to seek a berth on board one of the 50 or so older diesel trawlers, which now service and supply the North Sea gas and oil platforms. At the time of writing (May 1980), the number of mid-water trawlers actually fishing out of Lowestoft has dropped from 70 to 40, and the future is at best uncertain.

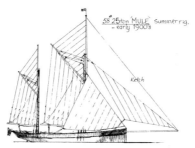

58 25ton MULE Summer rig. – early 1900's

Ketch

CHAPTER TWO

Haul The Trawl, Boys, Haul

The Days of Sail

'Once I was a schoolboy
And I stayed at home with ease.
Now I am a smacksman
I plough the raging seas.
I thought I'd like seafaring life,
But very soon I found
That it wasn't all plain sailing, boys,
Out on the fishing grounds'.
(Traditional − 'The Smacksman')

The craft on which the modern British trawling industry was founded were tall, powerfully canvased ketches, generally referred to as smacks. Smack is an old word which can be taken to mean any sailing fishing boat, but when the Lowestoft fishermen said 'smack' they referred specifically to a ketch or dandy rigged trawler over about 15 tons. The herring drifters were similarly rigged but always referred to in Lowestoft as 'luggers', a throwback to the much older dipping lug rig from which both types of fishing vessel were descended. Round in Brixham they often referred to their smacks as 'the big sloops', even though these too were ketch rigged. They never had been sloops, but were sometimes cutter rigged. Such are the pitfalls for the writer who tries to be too pedantic about correct nomenclature.

As these craft grew in numbers in Lowestoft from 1840's and 50's they brought with them a growth of allied industries: net manufacture, rope-making, fish-selling and handling, sailmaking and shipbuilding. Above all shipbuilding. The Lowestoft yards produced a string of fine vessels in the late 19th and early 20th century, taking note of what Kent and Devon had to offer in the way of smack design, but confident also in employing local knowledge to produce craft that would fish effectively and stand the worst weather that the North Sea could throw at them. Among the names still remembered are those of Samuel Sparham, Henry Reynolds, John Chambers and Sam Richards − quality builders all of them, who by their craftsmanship and integrity commanded the respect of boatowners and fishermen alike.

A smack of around 50 tons net cost about £850 ready for sea in the mid 1880s (exclusive of gear), and this was a good investment at a quarter the price of one of the new larger steam trawlers. Two decades later the cost had risen to about £1,450, but still showed a good return for local boatowners. Indeed, smacks were built for Lowestoft owners right into the 1920s, so faithfully did the port stick to sail. Of the other trawling ports only Brixham stayed with sail as long. Even with the considerable incidence of total loss, the average working life of a Lowestoft smack was something in the region of 15-20 years − and many of them went on a good deal longer than that. What made the vessels profitable was the fact that they were fishing local grounds for a specialist trade. The hauls of sole and plaice, taken on trips of about a week's duration, were always able to fetch a good price on the 'quality' market because the fish were landed in good condition. And

A wherry bringing ice to the sailing trawlers in Lowestoft dock. The smack in the middle of the photograph is *Inverlyon LT687* built in 1903 which later became one of the gun smacks. There is a good view of her rather crooked beam along the port side terminating in the iron hoop. Ahead of her is *John Browne LT327*, built in 1890 and on the extreme right *Nancy LT41*, built 1904.

not only this. Until the 1920s, Lowestoft had no real competition from steam trawlers. The smacks had it all their own way.

A typical Lowestoft boat of the late 19th/early 20th century was something in the region of 50 tons net, having an overall length of 70 feet or so, a beam of 18½ feet and a hold depth of nine feet. The total sail area was something in the region of 2,900 square feet which gave the smacks a top speed of around 12 knots. 'Sail better'n a yacht!' is a comment still made in Lowestoft about the trawling smacks.

The largest smacks working out of Lowestoft went up to about 68 tons net, while at the other end of the scale were the toshers, which were around the 30-35 ton mark. The latter were ten or a dozen feet shorter than their larger counterparts and they carried a crew of four, as opposed to five. Skipper, mate, third hand and cook was their complement, instead of skipper, mate, third hand, deckie and cook – though in the summer months both of them often reduced their crew number by dropping the cook. The change simply meant that his duties were simply taken up by one of the other crew members – tasks that he could reasonably cope with because the weather was so much easier than in winter. Many a Lowestoft boy began his fishing career as cook on board either a tosher or a smack, and many rose to become skippers of their own vessels.

The trawl beams varied in length between 30 and 50 feet, depending on the size of the boat working them, and they were some eight to nine inches in diameter. Hence a tosher would carry a beam of about 30-32 feet in length, while the largest smacks would have one of 48-50 feet. The beams were always carried on the port side of the Lowestoft boats and shot away on that side too, but for some reason many of the Yarmouth smacks opted for stowing and working their trawls to starboard. The wood used was either oak or elm, beech or ash, with the last-named variety the most popular. Lengths up to 30 or 35 feet were usually made in a single piece, but the longer beams consisted of two lengths roughly scarphed together, with the joint reinforced by iron bands around the beam. The beams were squared off at the ends to fit into sockets on the large, stirrup-shaped, iron trawl heads, which could weigh anything between two and three hundredweight each. Their function was to keep the top of the net some three or four feet above the sea bed to prevent fouling as the trawl was dragged along, and to compensate for the abrasive acton of sand and rock they were doubly thick along the bottom, or 'shoe' as it was known.

The net itself was a triangular, bag-shaped affair which was lashed to the trawl beam on its upper side and which had its deeply curved under-portion fastened to a thick, heavy ground-rope. This latter was further weighted down with lengths of chain that served also to achieve the sanding-up process so necessary to disturb flatfish and make them rise into the mouth of the advancing trawl. With the top of the net being fixed to the trawl beam and the bottom having a pronounced curve, a good deal of the trawl was already actually above the fish as they rose from the sea bed. Given the pattern of fish behaviour, which makes them usually swim into the tide, and remembering that the smacks always fished down-tide, it wasn't long before the various species found their way down into the cod end at the tail of the net. The meshes here were about one and a half inches square, as opposed to three inches or so over the rest of the net, and this is where the bulk of the catch lay – here and in the pockets, which lay just up-net from the cod end.

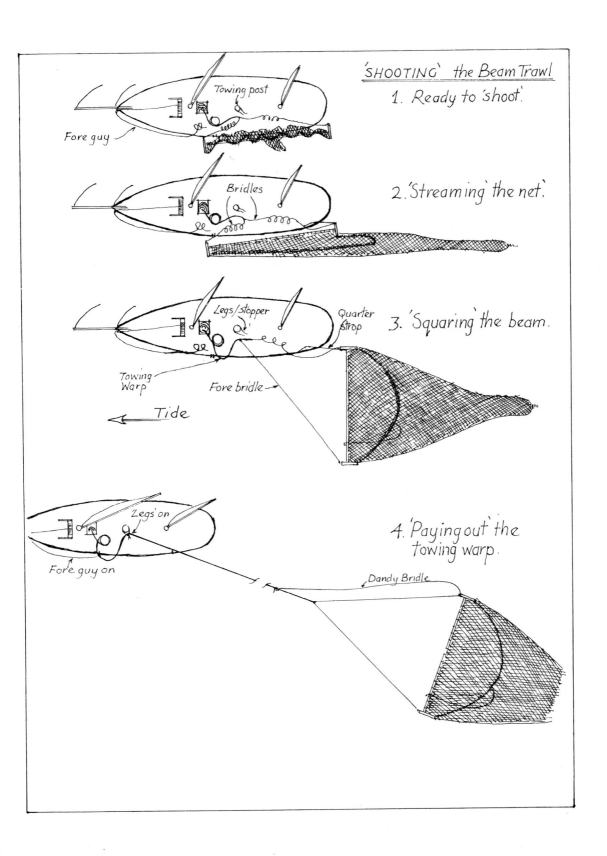

'SHOOTING' the Beam Trawl
1. Ready to 'shoot'.

Towing post

Fore guy

2. 'Streaming' the net.

Bridles

3. 'Squaring' the beam.

Legs/stopper

Quarter Strop

Towing Warp

Fore bridle

Tide

4. 'Paying out' the towing warp.

'Legs' on

Fore guy on

Dandy Bridle

The whole trawl was attached to the manilla warp (some 660 feet long and six or seven inches in circumference) by two rope bridles of around 90 feet in length which led from each trawl-head. There was a third bridle too (often made of wire) known as the dandy, which was slightly longer than the other ones and which was secured to the after trawl-head at one end and to the main warp at the other, just above where the two rope bridles joined. Its function was to bring in the after trawl-head when hauling, and there was another independent strop called the leach-line which drew in the heavy ground-rope. When towing along, the whole strain of the gear was taken by a double thickness of plaited manilla rope known as the stopper. This was doubled round the warp and then secured to a stout wooden upright called the towing post or dummy, which was set from the deck right down in the keelson of the boat. The stopper was of less breaking strain than the warp, so if the trawl came fast it would break instead of the main rope, thus saving the gear. When this happened, the strain was then thrown on to a rope guy, which was fixed to a winch up for'ad and which hung slack against the hull before becoming attached to the warp just abaft of where the latter ran out through the bulwarks. In taking the weight of the trawl through being secured to the warp, the guy would also cause the smack to come up safely head to wind.

The motive power for hauling trawls was provided by hand-capstans at one time, but during the 1870s steam capstans began to be used. During the mid 1880s William Garrood invented his famous cylindrical model, which was the first step in making the Beccles firm of Elliott and Garrood the foremost suppliers of steam capstans to the fishing industry.

Such machinery was important in lightening the work-load of the men on board, yet the life remained a hard and dangerous one. George Stock (born 1903), spent part of his early working life on board the Lowestoft sailing smacks and is one of those who still remembers:

'When I left school there were so many people out o' work that my father made me git a job at Colby's shipyard. He bound me apprentice shipwright. Well, directly the war finished, they went crash. So us apprentices went up t' the United Bus Company works. I wuz up there for a bit an' then they started standin' people orf, so I thought, "Blow this! I'm goin' t' sea." So I went in two or three smacks as cook. I went in the *Endeavour LT494*. There wuz a rum ow boy skipper o' her. He used t' run up the riggin' with a chopper if he got a split net, or if there wun't enough wind, or if he dint git n' fish. Yeah, he used t' go up there with a chopper, supposed t' be after God! You know, "Come down here an' I'll cut your hid orf?" Coo, he dint half used t' carry on. I remember bein' there once an' havin' t' take a reef in. You used t' hafta put yuh rope through the eyelets an' work along. Well, I stood up on the little boat, got on the main boom an got inta the sail. The ow boy let her shape up t' the wind an' that shook me out o' the sail. I went down on the thwarts o' the little boat an' I thought my life wuz gone. But you survived!

'Another smack I wuz in wuz along o' a bloke called Jack Dyer. Screamer they used t' call him. We lorst s' much gear that the owner give us a wire an' rope warp. Moostly, the warps used t' be all manilla, but there were these ones what were half an' half. Once you'd shot yuh gear, that'd be wire from the capstan so far out an' then rope the rest o' the way t' the net. Well, this bloomin' wire, you couldn't coil it. You try coilin' wire! That'll go which way it want to. An' one day that curled over me; that got my leg in a bight an' over

HAULING

-beam up

'Hoisting the bag'
- using mizzen tackle

pockets & flapper

quarter rope for hauling groundrope.
(leech line)

Towing warp

Dandy bridle

Bridles

iron wedges

TRAWL HEAD
LOWESTOFT
& RAMSGATE

square

batings

Cod end

flapper

wings

belly

BEAM TRAWL NET – COMPONENTS

GROUNDROPE

rounding

BRIXHAM

BARKING &
YARMOUTH

chafing pieces

ASSEMBLED PIECES

Beam – usually, Ash, Beech, Elm or Oak

Scarphed & iron bound for length – up to 50 ft.

the side I go! I wuz lucky, though; I managed t' grab hold of a rope that wuz trailin' through the scuppers. Mind yuh, I hetta carry on all the trip. They bound me leg up an' I hetta carry on, draggin' it about. That wuz on the *Clipper LT335*.

'I hed two brothers in smacks an' all. One wuz a skipper. He got his skipper's ticket when he wuz 20 an' he went skipper o' the *Helen May LT392*. She wuz a new smack, but they made him come out of her because he wun't 21 an' so he hetta go back mate another year. He wuz skipper o' the *Boy Percy LT90* for three or four year. She wuz one o' Warman's smacks. There were a couple o' them Warmans an' they owned a lot o' smacks between 'em. One had all his boats named after the Scottish lochs, but this other one, what my brother Bert sailed for, he had the *Boy Percy* an' the *Economy LT612* an' names like that. All their smacks used t' lay in the west end o' the trawl dock an' I used t' love gittin' down there when I wuz a boy. Durin' the war Bert wuz skipper o' the *Early Blossom LT6,* an armed gun-smack; an' my other brother, Spider, he wuz in the big *Children's Friend LT174,* another gun-smack. My gran'father had bin a big smack owner at one time, he hed a fleet o' 16, but what he done wrong wuz when he started the Low'stoft Steam Trawlin' Comp'ny. He went t' Aberdeen an' bought the *Greencastle* an' them two others, but he couldn't make a go of it. They were steamboats, but they worked beam trawls. Well, beam trawls wun't no good on steamboats down here, so he went bust over that job.

'You used t' catch more prime fish on the smacks, like soles an' that, 'cause they dint go over the ground too quick. Thass like if you wanted t' catch soles in a steam boat, you'd ease the ow gal down so she wuz just rollin' over. But a smack used t' go over the ground nice an' slow, an' t' me they'd catch more prime an' plaice. Mind yuh, you wanted a breeze – though there again, some o' the real good ow skippers could shoot in the summertime without a breeze. They'd just work the tide. The only thing wi' that is that you're liable t' put the trawl on its back. Then, when you hauled, you wun't git nothin'. Yeah, the trawl could turn right over. See, there wun't enough way on the boat when you were shootin'. You must have a bit o' way on a boat t' keep everything out tight. Try t' shoot in a calm, an' you could shoot away after a fashion, slowly pay away, but you did hafta watch out for the trawl goin' over on its back.

'Like I said, the beam trawl wuz good f' soles an' plaice, but not s' much for swimmin' fish. You've sin pictures o' the big trawl beams, I spect. You had the big ow iron heads with square bits on what the ends o' the trawl beam went in. Then there wuz the eye-bolts on the bottom where you shackled the ground-rope on. Course, yuh bridles ran out from the trawl heads as well. They'd run up t' meet the wrap an' there'd be links t' fix 'em inta the eye o' the warp. Even when you wun't movin', the beam would still keep the trawl open. Everything wuz fixed, yuh see. I remember when I wuz mate o' the *Brent LT98,* a steam trawler, we were workin' out on The Knoll an' we picked up an ow trawl head what looked like a penny-farthin' bike. I told my father when we come in an' he say, "I'll come down an' hev a look." He come round an' hed a look an' he say, "Thass one o' Durrant's fleet from Yarmouth. They tried that out in all their smacks." That wuz like a big wheel with a square at the top, where the trawl beam head went in, an' then that hed a thing run down with a little wheel at the bottom. My father told me that they were the only firm what had that – Durrant's o' Yarmouth. But that wuz years ago. You know, afore my time.

'The smacks' time out varied, but that'd usually be somethin' like seven, eight, nine, ten days. I mean, some of 'em would git becalmed summertime. Yes, that'd take you a long time t' git in summertime. That wuz like when you come up t' the buoys, when you used t' head f' the Stanford Channel. If you dint git in afore the tide turned, you'd hefta go away t' the nor'ad agin. Over go yuh anchor, an' you dun't heave it up till the next tide came along. That wuz the only way you could git in. But with a breeze o' course that wuz diff'rent. I mean, they used t' come in through the pier heads an' the bridge over the harbour would open up for 'em. They'd go right the way through, up t' where the Co-op Factory is, an' run right onta the mud there.

'When they wanted t' git from the Waveney Dock round t' the Trawl Dock, you used t' run out a heavin' line. That wun't very thick, but that wuz strong, an' you'd run it away from the boat. You'd see the deckie with it coiled up in the little boat, a-scullin' away like mad. He'd be runnin' that line from the smack t' one o' the moorin' bollards. Anyone'd put the bight over, an' then them on board would heave away on the ow capstan an' work their way round inta the trawl dock. They'd pull the smack round on the heavin' line. Yeah, they'd put that on the owd capstan an' pull themselves round. You'd hear them owd capstans a-goin' all over the show.

'Durin' the 1920s none o' the boats made much money. The smacks dint do much good either along o' the rest. A lot of 'em were gradually dyin' out. I mean, look at the big firms what went out o' business! Slater an' Barnard an' the rest. I mean, that wuz all right havin' smacks, but you'd gotta hev 'em caulked every now an' agin 'cause they're a wooden ship. Yes, they'd gotta go on the slip an' be caulked an' everything. An' then the owners, they'd all gotta hev their own yards an' they'd gotta hev their own sailmakers. You'd be surprised at the cost o' runnin' a smack, where in the old days, I mean, they'd git people t' do things f' next t' nothin. Well, thass my idea anyway.

'Now I got 7/6d a week when I wuz in the *Endeavour* an' that wun't a lot. The firm wuz up Commercial Road, an' strange t' say that'd bin my gran'father's store once, only two or three o' the small smack owners hed it then. There used t' be three crews work on tannin' the sails, only you dint use the cutch like you did for a drifter sail. The smacks' sails were red, not brown. What did we used t' call that? – barkin'! Thass right, barkin'. Barkin' sails. We had a big concrete yard an' we used t do 'em there. You used t' git the cakes o' horse grease an' put two or three o' them in with the water an' this red stuff (red lead). You hed a copper thing t' mix it all up in, then you used t' git buckets full of it. You hed big ow swabs t' work with an' you'd start at the top o' the sail an' work down. You'd hev yuh long sea boots on, yuh know.

'When you finished dressin' the sails, we used t' take ours t' the bottom o' Stevens Street. There used t' be a big field there, where the fair used t' go. We used t' lay them all out there, an' then the next day we'd go an' roll 'em up. Then they used t' write the LT an' the number in the sail, each side. I wuz only a boy then, like. I know we used t' hetta go down t' the "Eagle", in Tonning Street, t' git beer f' the ow boys. We used t' take these big stone jars down there, us younger ones, an' we used t' git pop an' beer. Cor, an summer time, that wuz lovely! Yeah, the shandy. An' the reg'lar ow boys, they'd hev the beer. The ow skippers used t' stand there an' watch yuh work, yuh know, an' you'd be a-sweatin' with this here drink, But, coo, that used t' go down lovely!'

It wasn't just a case of sweating in summertime for the smacksman. Even in the middle of winter the work on board was heavy enough to raise the body temperature. Ned Mullender (born 1896) talks about his experience of the hard slog of trawling in smacks before the First World War:

'When you shot yuh trawl, you paid the net over by hand. Then you used t' let the beam go. There'd be a man up for'ad an' he'd slip the lashin' on the for'ad trawl head an' let go. Then as the beam came round, the skipper would git hold o' the after lashin' what wuz on the other trawl head an' square the gear. As soon as the beam wuz square acrorss the stern an' parallel with the surface o' the water, he'd slip the lashin' an' away go the trawl. Away go the bridles too, what'd bin coiled down on the deck. Course, you'd got other things t' do an' all, like makin' yuh dandy fast on yuh warp just above the bridles, an' gittin' yuh guy onta the warp. Oh yeah. Then you'd gotta wrap yuh stopper round the warp an' hitch it t' the towin' post. Course, you'd let enough warp run out afore you did that. Once the stopper hed got the weight o' yuh gear, you'd take a couple or three turns o' the warp round the capstan ready f' haulin'. One end would be down the rope-room, yuh see; you never let it all run out.

'As soon so you were riddy for haulin', you'd undo the stopper an' let the warp run inta the roller-gang. Then you'd start t' haul. The capstan would wind in the warp an' the cook would be down below in the rope-room a-coilin' it. Then there'd be another man a-windin' in the guy on the windlass up for'ad. When yuh bridles come up, you'd take yuh dandy orf an' pass that aft, outside yuh riggin', round a roller-cleat an' up t' the capstan. Now you'd heave up till both the trawl heads were level an' then you'd lash 'em inta position, fore an' aft. Then you'd git yuh leach-line round the capstan an' heave up yuh ground-rope, an' once you'd done that you'd pull the rest o' the net up by hand. You'd pull yuh cod end up with a tackle on yuh mizzenm'st. Thass the reason yuh hed a mizzen mast what sloped forward. You'd git a guy round the cod, through the tackle an' run it t' the capstan an' heave up. The bag'd come just over the rail an' you'd slip the knot an' drop yuh fish down onta the deck.

'Now if you were fishin' the other side, starboard, you'd pass the gear round the stern. You shot the net an' the beam from the port side, but you'd pass the bridles round so when you let go o' the trawl, you'd let go on the starboard side. That wuz when you were workin' a weather tide, 'cause you'd shoot the trawl opposite side to it. You're heard 'em say I spect, there's bin a lee tide or a weather tide − well, you'd fish accordingly. Mind you, the net an' the beam allus went over the port side, but when you squared it up the bridles were passed round outside everything an' the warp wuz worked orf the starboard side.

'You towed the trawl with the tide an' as soon as the tide began t' slacken, you'd haul. That'd be about a four or five hour tow, but some of 'em used t' tow along all night. They'd work two tides an' pass the gear round the stern t' git the second one in. That'd be if they were acrorss on The Old Man's Ground − acrorss easterly on the Brown Ridges, nearly acrorss t' Holland. Yeah, you could tow along there all right an' you used t' git a lot o' these little ow plaice, about the size o' the palm o' yuh hand. You know, that wuz afore the regulation size come in. Tom Wright's titty ones, we used t' call 'em.

'A lot o' the top skippers used t' work down the Haisbro certain times o' the year, or up along from the Hinder an' the Rough Ground. When my father wuz in the *White Heather LT1013*, he wuz a-towin' along on the outside o' the Haisbro an' the ow smack come tight t' somethin'. Well, when they hauled they hed so much rooker in the bag that they couldn't heave em aboard! What they did wuz larnch the little boat − that wuz fine weather − unlace the trawl an' git a lot of 'em aboard in baskets. Then they laced the trawl up agin an' hove the rest on board. He come round the north end o' the Haisbro an' took the flood tide hoom. He only went out that mornin' an' he wuz back in Low'stoft harbour the next mornin' with 48 boxes o' thornyback rooker. About eight stone t' the box, I think. Course, the word got round an' a lot o' the boys went after 'em. A couple of 'em went aground! Yeah, an' all for about a quid a box!

'The length o' trip varied accordin' t' how yuh did. So did what yuh earnt. You might make £40 or £50; you might git only about £25 or £30. That all depended. You'd be at sea about a week. If you went t' sea o' Saturday − unless you caught a lot o' fish, in which case you'd come in an' land 'em − you'd probably come in o' Sunday mornin' an' land on Monday. I should think we hed somethin' like a couple o' hundred an' fifty smacks here afore the First World War. The trawl dock used t' be full. My father used t' moor up in the yacht basin winter-time, when the yachts wun't about. When you come in on a good blow, you used t' use a drogues'l sometimes. That wuz a cone-shaped canvas thing on a big round hoop an' you used t' hull that over the stern t' slow yuh up. That'd fill up wi' water an' act as a brake. You'd use that when there wuz strong easterly winds. You'd hoist a flag up on board an' they'd open the bridge for yuh. One ow Pakefield skipper come in one time on a bit of a blow an' ow Munnings, the berthin' master, say to him, "Shove her in the crick, skipper!" That wuz a place there near Morton's Factory. "Crick be buggered!" this ow boy say, "We're goin' this way t' bloody sunset!"

'You hed a crew o' five on the smacks; the toshers hed four. They were the under-tonnage ones − you know, about 25 tons or so. At one time skippers dint hafta have a ticket for them. They could just go. Some o' the smacks used t' go round t' the Westward, yuh know. They used t' go round t' Padstow after soles in the springtime. My father wuz round there in the 1880s. They used t' pack up the fish theirselves, the crew did, an' then the sellin' agent would send 'em orf t' London. Course, later on my father wuz a big man in Boardley's firm. He hed the *Crystal LT325*, he hed the *White Heather LT1013*, an' he hed the *Purple Heather LT249*. Oh yes, he did out an' out well in that firm.

'On the smacks the cook got 10/ − a week, then the deckhand would git about 12/6d. The third hand, he would git roughtly about 16/6d, an' the mate would draw about £1. The skipper, he'd git about 30 bob. This is afore the First World War, I'm talkin' about. The skipper an' the mate were paid on a share o' the catch, so much on the clear hundred after the boat's expenses had bin paid. The rest o' the crew were on a weekly wage, plus a little bit o' stockie bait. That wuz the money got from sellin stuff what wun't much value − gurnets an' things like that. My father used t' dole everyone out alike; he believed in equal shares in that matter. If the mate got 7/6d, which wuz a good stockie bait in the winter time, the cook got it as well.

'Now some o' the skippers used t' try an' keep moost of it for themselves. They hed the dolin' out of it so they used t' work a fiddle. One smack I went in, in the winter o' 1913, the skipper tried it on there. I went cook, an' when we come in the first trip I knew we had a good lot o' gurnets, so that shoulda meant some decent stockie. Well, when that come to it, the skipper give me sixpence, and a white hat − one o' these calico things with a stiffish brim. I say, "What's the hat for?" He say, "T' keep the sun orf yuh." In the winter this wuz! So then I say, "Well, what about this sixpence?" He say, "Thass yuh stockie. What'd you expect t' git?" Well, I knew with them gurnets that we shoulda hed praps five or six bob apiece. So then he say, "Aren't yuh satisfied?" I say, "No, I ent. Here yuh are, here's the sixpence back." Well, blast me if he dint go an' tairke it! So I dint hev nothin', only my hat.

'I hed another do wi' him an' all. He accused me o' usin' too much milk f' the tea. We only hed three tins o' condensed milk t' last a trip, where my father allus used t' allow half-a-dozen. An' not only that − this ow boy used t' sleep on top o' the grub locker, even though he wuz the only one what hed a key! Anyway, this here particular time, I'd made several brews while we were in the harbour so we runned out afore the end o' the trip. Well, when I went an' asked him for another tin, the arguments started! What'd I bin doin' with it an' all that sort o' thing. I told him I hent bin excessive an' what we used t' hev when I wuz along o' my father. He dint like that very much an' he got hold o' a rope's end an' wuz gorn t' draw me acrorss the behind with it. I say "Do you come an' try it!" I say, "If you strike me wi' that rope, I'll smite you wi' this!" An' I picked up a handspike. Well, bor, that mairde him dance back. Then my uncle, who wuz deckie there, stepped betwin us. He say, "Thass enough o' that. Let the boy alone." So the ow skiper walked away. That night we brought up here orf Pakefield in a flat-a-calm. We put the anchor over an' o' course I could hear some o' the lads a-playin' on the beach, so I thought I'd swim for it. I come up on deck wi' my clothes all tied t' the top o' my hid an' my uncle say t' me, "What're you gorn t' do!" I say, "Swim ashore." He say, "Dun't you be silly, boy. An' even if you do mairke it, you'll look all right walkin' hoom like that!" Well, I did as he said, an' I stayed on the boat another two trips till they got someone else as cook.

'The ow smacks used t' go well down on the North Sea, yuh know. Oh yes, they used t' go well down. Down onta the Oyster Bank, down there in the deep water orf Skillen Corner (Terschelling Island). Oh yes, you'd go a fair way from hoom. Oh yes. Some o' the ow skippers in the Short Blues, the Hewett boats, used t' go down a long, long way an' all. Some o' the Low'stoft ships used t' go fleetin' an' all, but not very many. No, not like they did out o' Yarmouth. Course, the beam trawlers you're got today, thass a different method altogether. They tow along on a boom orf the side o' the ship an' they use I dun't know how many o' these here chain titlers. But, o' course, they're got very powerful engines. Now I can't remember havin' a titler on in the ow smacks, though I think some of 'em did work just a little single titler on the bosom.

'I remember one smack there I wuz in once, the *Sincerity LT952*. She belonged t' my Uncle Wallace an' we went t' sea the day after Boxin' Day, 1913. While we were out, we carried the mast out of her, the foremast. We were jivin'; you know, runnin' afore the wind. Jivin' is where the sail go from one side t' the other. See, you're runnin' afore the wind an' you're got yuh sail slackened out as far as what yuh rope would let it. Yuh main sheet, I'm talkin' about. Well, if the sail suddenly got a gust o' wind, an the ow man let her sheer a bit, she'd praps give a roll an' the boom'd swing right acrorss t' the other side.

Well, thass what happened with us, 'cept that the boat dint roll enough an the force o' the boom swingin' acrorss wuz enough t' carry the mast out of her. That brooke about 20 feet above the deck, so o' course we hetta chop away the riggin' an' let the mast hang on the forestay, what come up from the stem. An' thass how we rid out a gairle o' wind, like that! Talk about roll! The water wuz comin' over the rail one side an' then over the other. I thought t' m'self, "This is all right. Thass what come o' bein' in too much of a hurry." See, we'd bin sailin' hoom flat out so's we'd could catch the mornin' market.

'We were messin' about a good bit o' the night with this here broken mast, but about nine o' clock in the mornin' away come a drifter-trawler, the *Vesper Star LT94*. Ow Walter Goslin' wuz skipper. "Whass up, John?" he say t' my father, 'cause my father wuz skipper, yuh see, an he allus called him "John". "That look as if you'll hetta pull us in, Walter," say my father. So thass what happened. We come in on a tow. We hetta saw away the forestay with a hacksaw 'cause that wuz made o' wire; an' once we'd done that, away go the lot. Mind yuh, that dint break the borsprit or nothin'. We just run that in an' cut the rest away. When we got in, we laid there for a bit an' hed a new mast put in. We were all right 'cause that wuz an insurance job. Once that wuz done, out we go agin – an', blimey, if we dint go an' break the mizzenm'st next trip! Well, that dint break alt'gether, but that split so we dussn't put too much weight on it. So hoom we hetta come wi' that. We got some rope round it an' bound it up; drove wedges into it so that would hold up. Yeah, thass what we did. We even managed t' finish the trip.

'You wanted a breeze f' smackin'. Not necessarily a reef breeze, though that wuz all right too, but about three or four on the Beaufort Scale. That wuz nice workin' weather 'cause you could git where you wanted to. In fine weather you hetta tide about a lot, but with a nice tops'l breeze you were well away. You know, I'm talkin' about yuh winter tops'l. That wuz a small one, a three-cornered one. Triangular you'd call it. Yuh summer tops'l wuz bigger an' that hed a garf on it. You'd change yuh sails f' the winter; you'd hev newer ones. See, you used yuh old sails in the summer t' use 'em up an' the weather wun't s' bad then. But f' the winter you hetta hev the best. A set o' sails corst some money, yuh know. Course, I mean, they were made o' canvas an' they were all hand-sewn. My Uncle Wallace hed four smacks at one time an' he hed his own sailmaker, a chap b' the name o' Harry Reynolds. He used t' make the sails f' my uncle's boats up on the net store.

'That wuz very rare you towed round the clock on a smack. You'd git in as much towin' as you could, but you wun't go all round. You dint lose no more hauls than what you were forced to, but you wun't go all round. Course, that all depended what ground you were on an' what conditions were like as t' how you worked. I mean, we hev hed two hauls on a tide. On about a six hour tide we'd git two hauls in. What you'd do wuz come the length o' yuh ground an' then haul. Then you'd sail her up t' wind'ard an' come back down agin on the tide, see, till the tide wuz done. An' thass the way you kept on yuh ground. Once the tide wuz done, you'd go the other way back. An' if you couldn't git back, well, you'd sail on an' make all lee tides, yuh see. The numbers o' hauls you managed in 24 hours would vary a bit. You might git four in; you might git as many as six if things were right an' you were on a lot o' fish.

'Once you'd hauled, you'd gut yuh fish on deck, wash 'em an then send 'em down below. You'd wash yuh soles in a tub with yuh fingers, but you hed what were called fish-bags f' the rest o' the stuff. They were braided up out o' net an' they hed a rope round the top what would pull up tight. You'd fill them up wi' fish, pull the rope up right an' throw 'em over the side, an the force o' the ship goin' along used t' sluish 'em through. Then you used t' put 'em in baskets, let 'em drain an' then send 'em down below. Like I said, the soles were allus washed out in a tub. The mate used t' do them. That wuz his job. The fish were kept in pounds down below. You'd put, say, three or four baskets o ice in the bottom o' the pound, spread it out an' then you'd put so many baskets o' fish on top. Then you'd put another layer o' ice down an' another few baskets o' fish on top. Thass how you'd build 'em up. You carried the ice down in yuh fish-room. You hed an ice-box there, a proper built one. That wuz tin-foiled or zinc-foiled on the inside an' there wuz a door into it so you could git the ice out. You dint git a great quantity o' fish in a smack. I can't really remember what a typical catch'd be. I should think a goodish one woulda bin praps 70 or 80 boxes. I remember one o' the smacks makin' £74 one Chris'mas trip an' that wuz very good, but I can't remember what boat it wuz.

'The livin' conditions were all right on board them ow boats, yuh know. Some o' the skippers were a bit mean wi' the grub, but there wuz allus enough t' eat. You wun't ha' starved. There used t' be four bunks down in the cabin an' the cook used t' sleep on what they called the after-locker. He'd be curled up round the mizzenm'st. I believe some o the smacks did hev five bunks, but the ones I went on hed four. That wuz all right sleepin' behind the mast – especially if you wuz a-towin' along, 'cause you'd put yuh head up t' the weather side an' lay there nice an' easy. But if you were tackin' about durin' the night an' tryin' t' git t' wind'ard anywhere, well, one time yuh head would be up an' then the next time that'd be down. Yeah, the blood would run inta yuh head then! If you were tackin' t' wind'ard, you'd miss a tide somewhere t' git back on yuh grounds, an' consequences wuz if you were tackin for about an hour you'd be layin down head first very nearly!

'You'd be up the Hinder in January on the smacks; up orf Harwich, only further out acrorss the other side. You'd git plaice, soles, rooker an' whitins an' that sort o' stuff. Durin' the summer you'd be down on the Haisbro' an' the Long Shoal an' Hammond's Knoll. You'd git rooker an soles an that sort o' thing there. In the autumn you'd be playin' about where you could git, out o the way o' the drifters. You'd be in amongst the Long Shoals, where they couldn't come, an' you'd be gittin', codlin', whitin' an' thornyback rooker. Now when you went up t' the south'ard, you'd be gittin' smooth-back rooker, what they called the blond rooker, an' they used t' make more money than the thornybacks. That wuz orf the Hinder an' acrorss t' the Hook o' Holland. You never got much cod on the smacks, a box or two maybe, but nothin' big. No, soles an' plaice were the main things, soles an' plaice.

'Thass difficult t' say how good beam trawls were, but I should say that a beam trawl of the same size as an otter trawl would catch more flatfish. Without titlers, I'm talkin' about. Mind yuh, an otter trawl would catch more long fish; you know, whitins an' cod stuff. You used t' hev dangles on a smack's foot-rope. They were round rings, about six inches in diameter, linked t'gether by a chain about a foot long. You'd thread them all along the foot-rope an' they'd be fixed t' the trawl-heads at each end. They used t' weight the ground-rope down an' cause the sandin' up you needed t' catch plaice an' soles. Some o' the ow skippers just used t' bind lengths o' chain round 'cause that wuz cheaper. Oh, there were all sorts o' ways o' doin' things. Every man hed his own idea'.

The last remark is probably the only statement that can be safely made about trawling in general which won't bring either immediate contradiction or denial. Herbert Doy (born 1900) now recalls his early days on the sailing smacks, some of them spent in the company of Ned Mullender's father:

'I started smackin' when I left school, in the *Dragoon LT242,* a tosher. I went cook in her. That wuz winter time, November, when I finished school. There wuz only four hands in her winter time an three hands in the summer. In winter time the skipper, the mate an the third hand hetta pay the cook's money if they wanted a cook, an' in the summer time the skipper'd usually cook. He'd do that betwin seven in the mornin' an' two in the afternoon. Then they'd shift watches. Thass when they'd hev their dinner, two o' clock in the afternoon. The skipper'd come on watch at seven an' they'd hev dinner at two. Then the mate an' the third hand would take the afternoon an' the night watches. Thass how they worked it. I got nine shillins a week when I started, an' a shillin' stocker bait.

'We used t' go acrorss easterly, about 40 mile from the Dutch coast. That wuz a smooth bit o' ground out there an sometimes you could git two tides in. You'd git a lee tide an' a weather tide. On a lee tide, if there wun't much wind, you could drop yuh gear down, but on a weather tide you'd want a breeze. A reef breeze is a nice breeze for a sailin' smack. Thass about force four or five. When you hed that, you could sometimes git two tides in then. When you got a weather tide, you'd gotta take all yuh ropes right round aft an' bring 'em up on the other side. But you allus dropped yuh gear orf the port side. You'd got a spare beam laid on the deck starboard side an' yuh little boat wuz there an' all.

'We used t' catch plaice, soles, rooker, gurnets an' latchets. Latchets are like a gurnet, only bigger. I wish I could git some now. Thass years since I hed a nice baked latchet. We used t' git brill an turbot as well, all that sort o' thing. You'd even git a stray cod or two. That wuz mixed fishin' on the smacks an you'd allus git a good lot o' small plaice over near the Dutch coast. When we worked The Knoll, in the gut o' The Knoll, which is about two miles inside the lightship, we used t' pull up the truck on the mainsail an' the mizzen, lower the fores'l an' just hev the jib out. The truck wuz the bottom part o' the sail near the mast. There wuz a hoop on there, with a tackle, an' you used t' pull up on that t' shorten yuh sail. That all depended on the wind how much sail you hed. When you were towin' on a weather tide, youd gotta hev plenty o' sail. Yes, you'd hev everything up then 'cause you're really gotta pull. But on a lee tide you could just drop down wi' the gear. You dun't want half s' much sailin' on a lee tide.

'You'd tow along till the slack o' the tide. About five or six hours that'd be. The warp'd begin t' shake when the tide slacked. That wuz because the beam wuz playin' about on account o' the turn o' the tide. You'd haul then. That wuz hard work too − plenty o' pullin'. I hetta go down an' coil that bloody gret warp. That wuz nearly as thick as my leg! You'd coil that up an there wuz the two bridles shackled on it. You'd unshackle the after one from orf the warp an' then you'd run up an' run aft an' turn the donkey on t' pump the boiler up 'cause all the steam'd be gone. The water wuz gone out o' the boiler with the capstan windin' away, so you'd gotta go an' fire up an' pump the boiler up again.

'I went on the big smacks as well as the toshers. You hed about a 38 or 40 foot beam there instead o' 32. I wuz in the ow *Jessamine LT93* durin' the 1914 War. They were all old blokes in the crew, bar me. I wuz only a boy. Ow Spencer Broom wuz skipper an' one day a bloody mine come against our after trawl hid. You oughta seen 'em runnin' with a boathook t' push it away! That wuz right against our trawl hid an' they were shovin' it away with a boathook. They were nice ow boys on that boat. Ow Spencer Broom wuz a church-goin bloke. We come in the harbour one day an ow Wincup, our owner, come down. He say t' Spencer, "Mornin', Broom." Ow Spencer say, "I've got a handle t' my name, guvnor." So ow Wincup say, "Mornin', Broom handle." Ha, ha, ha, ha, ha! Thass a lot o' years ago, but I'll never f'git that. "Mornin' Broom handle."

'I went down t' Padstow one year along o' ow Ned Mullender. You hetta work bobbins on yuh ground-rope round there. Course, normally you just hed yuh dangles, but round there you used t' work small wooden bobbins. Beads we used t' call 'em. We used t' thread 'em along the ground-rope 'cause that wuz such rough ground, all slate. Oh, that wuz terrible. That'd cut a ground-rope t' pieces if you dint hev these beads. You'd use 'em out here sometimes on some o' the rougher bits o' ground. Places like down outside the Cromer Knoll. We used t' git the boat southed down there if we could. You know, workin' due south so you got a little bit o' clear ground.

'You hed false bellies on yuh cod end t' stop it chafin'; old bits o' net or cow hides were what they used. Course, a beam trawl's net hev pockets on above the cod end. They're on both sides an' thass where the soles sed t' git in. That wuz a good way o' fishin', a beam trawl. I mean, you could even work acrorss on the Broad Fourteens. Thass right acrorss easterly an' thass all banks an' ridges. You could work a beam trawl on there, where otter gear would be all over the place. Yeah, yuh doors would be up an' down on them banks, a-twistin' and' turnin' all over the place. Well, you'd soon rip a net like that, but the ow trawl hids would stand it. Mind yuh, you'd sometimes break yuh beams. Oh, you can break a beam easy. If you wun't smart enough haulin' when the tide changed, the whole lot could turn right over on its back. That'd tairke all yuh bloody hidline right away orf yuh beam an' break the beam as well. You hed a right ow job on then. You used t' hetta shove iron bands around the beam t' fix it up so that'd hold.

'You used t' gut yuh fish on deck. Then you'd swill 'em round in baskets in a big ow tub t' wash 'em. Soles hetta hev all the blood nipped out o' them an' be paired orf afore they were prop'ly ready. They were like gold dust t' them ow smacksmen. They used t' ice 'em away as if they were worth a million pound! The skipper an' mate used t' look after them an' they were iced down sep'rate from the other stuff. You hed yuh fish pounds down in the hold on each side o' yuh ice-locker an' yuh fish would be put away in them. There'd be a layer o' ice, then a layer o' fish on top, then another layer o' ice an' so on. The rough stuff you'd chuck anywhere. You know, rooker an' that. Rooker! They dint pay no regard t' that sort o' thing! They just used t' chuck it in the bottom o' the pound an' make a ground tier. Just shove a little ice in the bottom o' the pound an' then chuck the rooker on that. Then some more ice, an' then you'd start t' put the plaice an' that on top. One thing you hetta watch wuz not lettin' a lot o' draught git inta the fish 'cause draught'll soon cut the ice away, melt it right out. Thass why you shipped yuh boards up the front o' the pounds, t' keep the draught out.

'The skipper used t' look after the fish when I wuz on the smacks, an' any small rooker wuz his stockie bait. If you were gittin' any quantity o' small rooker, you'd think o' the ow skipper. That wuz his perks. The crew got gurnets an' latchets an' weevers f' their stockie. They used t' sell 'em in heaps on the floor o' the market. A shillin' a heap! A basket o' gurnets f' a shillin'. That used t' be yuh stockie bait, layin' there on the floor o' the market, on the ground. On a tosher all the crew were paid on a share − all bar the cook. But in a big smack only the skipper an' mate were on a share. The third hand, deckie an' cook got a weekly wage there. The sharemen got allowance money while they were on a voyage, but that wuz deducted when they settled. They got a weekly allotment which wuz took orf their earnins. That mighta bin suffin like a quid a week, an' when that come t' the end o' the voyage that wuz all taken orf in the settlin'. The pay-out was made after all the runnin' expenses had bin allowed for. You know, food an' coal an' stores an' all that. A voyage used t' be about five or six trips an' a trip used t' be about seven or eight days.

'Some o' them ow skippers used t' be very fussy about the way their ships looked. The poor ow deckie use t' be on his hands an' knees comin hoom, scrubbin' down the decks wi' dorgfish an' nurses' skins. They were right rough, yuh know, an' he used t' hetta go all round the bulwarks wi' them. Then he used t' hetta clean the sidelights an' all. They used t' be kept down the cabin when they wun't bein' used an' he hetta polish them so you could see t' shave in 'em. The cook's job wuz t' clean the compass, take that out an' clean it. That wuz just the fore side o' yuh mizzen, so you could steer along an' look down at it. Oh, they used t' think the world o' them boats, the ow skippers.

'You used t' look round, yuh know, an' see if a smack hed got brass on the tiller. If that had, then you knew the skipper carried jam. Yes, thass what you used t' say − "She's got brass on the tiller. She must carry jam!" Course, you know, some o' them were funny tight about grub. You were only allowed about four tins o' condensed milk a trip. An' some of 'em used t' lay agin the biscuit locker. Thass the truth, that is! Yeah, they wun't git in their bunks; they'd lay on top o' the grub locker with a board shipped in in front of 'em so they dint roll out. There used t' be bread one side an' biscuits the other. They were the hard tack biscuits, the 25-holers. I've seen weevils comin' out o' them! Yes, but you dint pay no regard t' that. You just used t' shairke 'em out, soak the biscuits in salt water an' stick em in the oven f' a while. Oh, they were lovely like that.

'You used t' wash all yuh vegetables in sea-water. The only fresh water on board wuz f' drinkin'. You couldn't hev a wash once above a moonshine; you just hetta rinse yuh face over in salt water. Oh, you mustn't go drawin' on yuh freshwater tank or that wun't last the trip out. You never even hed a tank f' the boiler; that wuz salt water an' all. But, coo t' heck, if that dint fur up somethin' quick! Do yuh know, I've bin so short o' fresh water that I've gone down t' the ice-locker, got some ice out an' melted that down! That I hev. You dint hetta come on deck t' git inta the fish-room. There wuz an alleyway from where the boiler wuz through inta the hold, so I used t' nip through there an' git some ice out.

'The boiler wuz an upright one, like a marine boiler. The third hand used t' clean that out. Yeah, he used t' look after that, but the cook used t' hetta clean out the ashes from the fire. He used t' do that early mornins. You'd keep the fire in, but you'd hev a kind o' rake thing an' you used t' rake out the ash an' clinker from under the gratings. Yuh coal bunker used t' carry about half a ton. A bunker lid used t' lift orf on deck an' they'd shove you about half a ton in. You used t' keep an ow shoe in the boiler. A shoe-kettle. You used t' shove that in the furnace, an' you'd hev a lashin' on the handle an' hook that up inta this here bunker lid. That wuz f' the tea, the shoe, an' that wuz in the fire all day. That used t' last. You'd warm it up after that got cold. You dint git fresh tea every time afore you hauled. They couldn't afford that! Nor yit at every meal. No, you'd mairke one lot, tairke what yuh wanted an' then top it up f' the next turn-out.

'That wuz all wholesome food you hed on board. Blast, that time I wuz tellin' you about, when that there mine got caught aginst the trawl hid, I'd mairde a plum duff. That wuz Sunday an' I'd done plum duff an' treacle. They dint want n' bloody plum duff, though; they were all arter the mine, a-shovin' on it away. You could see the ow spikes on it an' there they were, a-shovin' it away with a boathook! You used t' hev fried fish f' breakfast an', cor, couldn't they eat! You'd hefta heap up a big deep platter wi' small plaice an' small whitins, all small stuff, an' they'd see that away. If you'd got a good deckie, he'd clean some of it for yuh an' hold that over till next day. That'd go on ice so you allus hed some riddy. You used t' mix up mustard, salt water an' vinegar in a big ow stone jar an' thass what you used t' hev on yuh fish. North Sea sauce they used t' call it. You'd hev two jars o' that – one t' use an' one riddy f' the next trip 'cause that hed t' be a week old t' be right.

'There wuz several smacks workin' out o' Low'stoft right up till the last war. One o' the last belonged t' Fred Moxey, the *Telesia LT1155*. He took her t' the Spithead Revue in 1935 f' the Silver Jubilee. She'd bin one o' the armed smacks in the First War. Course, a lot o' the Low'stoft boats were round at Padstow then. Thass when I wuz round there along o' ow Ned Mullender. Mind yuh, there were U-boats round there an' they sunk one or two smacks. One boat, the *Fleurette LT312,* she hed a submarine chase her an' she run aground. Yeah, she stood up on them rocks f' ages down there at Padstow. The poor devils on board got inta the little boat an' they were all lorst. If they'd a-stopped on board, they'd a-bin all right. Poor little ow Chalks Westgate wuz lorst in her. They asked him t' go as cook f' one trip 'cause his own boat wuz hevin' a refit. What about that f' a bit o' bad luck? One o' the smacks I wuz in round there, she never come home t' Low'stoft 'cause she wuz lorst round there. That wuz the *Young Clifford LT498*. She wuz a bran' new smack when I wuz in her along o' ow Ned. Everyone's heard o' the *Boy Clifford LT1202* here in Low'stoft, but not many ha' heard o' the *Young Clifford*. We took the *Olive LT299* after we'd bin in her, then I left Ned t' go steam trawlin' in the *Cuckoo M208* along o' Ted Chilvers. The smacks dint go round t' Padstow after the war, but the motor ones did. When they started convertin' 'em t' diesel, they used t' go round.

'That wuz lovely on board them ow smacks. You git a pair o' reefs down an' they lay as snug as a duck. Rye-built ships were the best ones. They'd stand the weather more'n what a Brickie would, a Brixham smack. Brickies wun't s' stiff a build; they wun't as strong as a Rye boat. Where you'd want one reef in a Rye smack, in a Brickie you'd want two.

The massive tiller of a sailing smack. In heavy weather it would be rigged
with tackles and took two men to handle. Very few had wheels for steering.
One exception was the *Lustre* mentioned by Bill Jarvis in this chapter.

Course, they built smacks here in Low'stoft an' all, but they wun't as good as the Rye boats neither. I wuz in the *Corona LT165* along o' Snowy Gooch an' we did 13 mile an hour comin' up The Roads in her! Till we got t' Yarmouth. We were comin' from Cromer an' we hed a wind. Soon as we got t' Yarmouth, we lorst the wind. But we were comin' past these little ow coasters there at one time; that we were. Course, we hed all the sails up t' do it an' the spinnaker out as well.

'You used t' carry all sorts o' sails on board. There wuz a big locker up for'ad, under where the windlass wuz, an' they were kept in there. There wuz a big hatch over it an' you used t' lift that orf an' go down. You'd hev yuh spinnaker, yuh nine-clorth jib, yuh seven-clorth jib, yuh five-clorth jib an' a storm jib. The spinnaker wuz the biggest an' the storm-jib the smallest. They all hed numbers on 'em so you knew which wuz which when you were gittin' 'em out o' the locker. You used t' work a stays'l as well; that'd be fore-side o' yuh mizzen. Then you might hev a tow-fores'l, what we used t' call a tow-dinger. That wuz a gret triangular thing an' that'd run from yuh fore riggin' right down aft. An' then you hed yuh tops'ls. The ordinary ones were triangular, but the biguns hed yards on 'em. Jiggers we used t' call them.

'When you were goin' along with any wind, you dint wanta steer them ow boats. They'd steer theirselves. You just used t' lash yuh tiller below the mizzen, pull yuh fores'l t' wind'ard an' she'd go along like a duck. Course, that wun't when you were towin' along; that wuz just when you were layin' about the sea, just keepin' her steady. Sometimes there wuz s' much weather you hetta hev tackles on yuh tiller t' steer with. Thass when there wuz too much wind an' too much weight on the tiller. You'd pull up on these tackles an' then you could git her t' answer. Course, you allus stand t' wind'ard when you steer an' the only bit o' cover you used t' hev, the only bit o' shelter, wuz a bloody ow bit o' canvas fixed up in yuh mizzen riggin'. A dodger they used t' call it.

'The skipper used t' take the tiller when you were stowin' up an' gittin riddy t' shoot the trawl, but that'd be the mate an' third hand an' deckie the rest o' the time. The deckie hed the mornin' watch, seven till two, alone, on deck alone. Then the mate or the third hand would take over from him, two in the afternoon till seven at night. An' then the next one'd be seven at night till one in the mornin'. Then one o' clock in the mornin' till seven. The deckie allus hed the daylight watch, the mate an' the third hand the night ones. You'd be towin' along, but only one'd be on watch. Course, everybody would be up an' about t' haul an' gut.

'You could git a nice sleep on them ow smacks, barrin' if that happened t' come on a gairle o' wind. Then you'd gotta turn out an' reef. You used t' jump inta that bloody mains'l, a-reevin' the ropes through the holes, an' that'd be throwin' water all over yuh. The skipper would be at the tiller then, seein' she dint luff, 'cause if she did luff up she'd throw yuh out o' the sail. You hetta watch what you were doin' on board them. You could git hit by the bloody boom when that go acrorss. Or by the main sheet. The block on that wuz a bloody gret thing, as big as the screen o' that television! Mind yuh, you never got no rollin' on the ow smacks. They'd go up an' down, but they wun't roll. They'd lay over like a yacht. Thass the reason you used t' git messed up wi' fryin' stuff sometimes — the fat all used t' lay in one side o' the pan! Mind yuh, they kept dry on deck; you dint ship half the water on them what you did on a steam trawler. No, you could go about their decks anyhow. You'd git a little spume drift, but that wuz all. That wuz just like someone blowin' little bits o' water over yuh. You dint pay no regard t' that.

'When you got a third reef, though, you used t' hetta watch it. Thass a gairle, third reef. You hev yuh storm-jib out then. That kept the boat's hid t' the sea. You'd gotta hev that; the mizzen wun't no good to yuh then. An' if you jived over, you could take a bloomin' mast right out! If anyone wuz a bit awkward on the tiller when you were runnin' afore the wind, an' yuh sail happened t' jive over, that'd snap yuh bloody mast just like that! An' even when yuh did know what you were doin', things could still happen. You could git clean swept. That happened t' one or two of 'em out here in the North Sea. They got blew down as far as the Dorgger Bank. Young Randle Harmer an' his crew did, in the *Our Merit LT535*. A gale blew all their sails away an' they finished up down on the Dorgger. Still, they did manage t' pick 'em up an' git 'em hoom.

'They were good blokes, the ow smacksmen. They used t' weigh the weather up to a bloomin' tee. They'd work the glass, yuh know. They'd see a blow a-comin' an' reef all riddy for it. "Right, we'll shove a reef in afore dark." Oh, they were good ow boys. They could come straight t' Low'stoft an' all. Yes, they could fetch Low'stoft all right! They could tack a boat in t' the Low'stoft piers. I've come in here more'n once an' bin castin' the lead all the time. You'd sing out what water you'd got — "Seven fathom!" "All right," they'd say. "We know where we are. Draw near the buoys." They used t' know all right. Keep her up when you see a bloody buoy. Thick weather an' all! They used t' git 'em in just the same. Thass an awkward harbour t' come into an' all, Low'stoft is — specially when you git the wind from the east. Mind yuh, thass worse now than what it used t' be 'cause there's that bar run right acrorss now. You used t' be able t' come in right from the East Newcombe at one time. An' once you got in, some o' them ow boys used t' bring the boat up inta the wind an' stop dead. That wuz if the wind wuz t' the south'ard. You'd be runnin' in round the dump-hids t' the harbour beach an' the ow skipper'd shove the tiller down, bring her right up hid t' wind, an' she'd stop dead. Oh, they were clever ow people, they were. Iron men in wooden ships. Thass the truth, that is! They were hard ow boys. They hed t' be.'

Bill Jarvis (born 1906), known as 'Winky' to all his friends, was a herring fisherman for much of his working life, but did some time on the Lowestoft smacks during the late 1920s and early 30s. Things were slack then in his home town, Gorleston, and so he shifted south (as others of his contemporaries did) to get a berth on board a sailing trawler:

'I wuz out o' work at the time an' a couple of us come over here t' Low'stoft t' try an' git somethin'. I see a notice up — "Cook wanted in a smack" — so I went inside an' that wuz Moore's orffice. Now, my mother hed bin at school wi' him in Loddon when they were kids, an' there I wuz arter a job in his firm. The man in the orffice say t' me, "Be down in the mornin'." So thass what I did, an' I shipped aboard the *Spray R170*. She wuz a Ramsgate tosher an' there wuz only three of yuh aboard her — the skipper, the mate an' me. The ow skipper lived in Golstun; his name wuz Harry Summers. When we got t' sea, he say t' me, "Put the ow shoe on, boy." That wuz a long thing wi' like a saucepan on the end an' you used t' shove that in the boiler t' boil yuh water up. Then you'd brew yuh tea up in the kittle. Some of 'em even used t' brew up in the shoe. Ow Harry Summers say t' me, "Hev yuh done any cookin' before?" I say, "No." He say, "Well I'll just show

Bluebell LT752 in Lowestoft trawl dock in the 1920's. The powerful square
stern and low freeboard aft were essential features of sailing smack design.
Note the lower landing stage to the quay on the right. It was built this way
to facilitate the landing of fish from the boats. The two men on the right are
wearing the typical tan calico jumpers and leather seaboots which were the
common dress of smacksmen of the period.

yuh.'' An' he showed me how t' mairke a duff an' that sort o' thing.

'Next thing he say t' me wuz, ''Dun't use too much grub, boy.'' He say, ''We're only allowed a pound o' cheese, two jars o' jam, four bits o' meat, a stoon o' bread an' a stoon o' flour t' last the trip. An' thass a ten day trip.'' Well, we were out 16 days in the finish! We got blown all over the North Sea. When we got t' Yarmouth, the tug come out for us. The ow *United Services* come out an' towed us in. The sails were all torn in ribbons. That wuz a bad gale, yuh see. Anyway, the ow skipper say, ''I'll see yuh in the mornin', boy.'' I say, ''All right.'' Course, I wuz hoom like a shot out of a gun! An' then I went down next mornin' t' help git the fish out. Just as I wuz goin' t' git aboard, blast if I dint slip an' fall over on a moorin' bollard! I brooke m' wrist, so o' course I couldn't go t' sea for a bit.

'When you shot the trawl, you used t' hook yuh two bridles onta the trawl hids an' git the net an' the beam over the side. You'd let that all go aft an' then you'd tow along on a gret big warp. You'd tow about three an' a half hours, praps four. All depended on what you were gittin', yuh see. The warp hetta be the right tightness, yuh know. Some o' the ow skippers used t' lay their faces on the warp an' feel the vibrations. They'd know when that wuz just right. When you come t' haul, you'd turn out an' the ow skipper'd say, ''All right, boy, just nip down an' mairke the tea.'' So you'd go down an' do that, an' then you'd all hev some afore you went t' work. My job wuz t' git down in the rope-locker an' coil the warp. Cor, that wuz somethin' thick! An' the first time I done it I wuz a-goin' round an' round; I dint know where I wuz. I howled. That I did. ''Never you mind, boy''. the ow skipper say. ''Pull down. Keep a-coilin'.''

'After I'd bin in the *Spray,* I went in one o' Moore's big smacks, the *Lustre LT823.* There wuz five in the crew there. The skipper's name wuz Green an' he wuz quite a nice ow boy. Well, they were all nice ow cocks, but there were rough. You know, they dint hev no sympathy wi' yuh. She wuz a good boat, though. Very dry. The only thing wuz that her boiler wuz different t' the other one, so I hetta git used t' that. They were upright boilers on the smacks an' they'd got a pump on the side where you pump out yuh billiges an' that. You used t' cook yuh meat underneath the stove, yuh cookin' stove. You'd put it in a bakin' tin, put yuh gravy an' onions an' everything in another tin over the top, an' then put 'em both underneath the stove. There wuz no table in the cabin, yuh know. You used t' set down there wi' yuh dinner on yuh lap. An' if the boat rolled, away went yuh dinner! Ha, ha, ha, ha, ha! Kiddies nowadays dun't know nothin', do they? The grub wuz all right on board, though, 'cept they used t' watch how much yuh ate. Yis, you mustn't eat too much! The ow skipper would soon tell yuh about it. ''All right, boy, we may be out here a long while. Dun't eat too much.'' Well, o' course him an' the mate, they'd git charged for the grub out o' what they earnt so they used t' keep everybody down.

'We used t' go fishin' out on the North Sea. Orf Cromer an' all round there, you know, but well out. Yis, you might run down a hundred mile time you'd done; the wind'd blow yuh all over the sea. Course, you used t' set yuh sails accordin' t' the wind an' you used t' hetta heave all them up. An' there wun't no wheelhouse on board! You hed what they called a dodger aginst where you steered. That wuz a big piece o' canvas fixed up in the mizzen riggin'. Moost o' the boats hed tillers, but the *Lustre* hed a wheel. Yeah, that she did. The ow *Spray* hed a tiller, an' if you dint hold that prop'ly she'd shoot yuh right acrorss the other side o' the deck.

'When you hauled on board them, you used t' pull yuh cod end up an' drop yuh fish on the deck. You used t' hev boards what went acrorss the deck, pound boards, an' they'd mairke a sort o' compartment f' yuh fish t' drop into. There'd be a tub there for yuh t' wash 'em in, an' then after they'd bin gutted an' washed they'd be iced away down in the fish-room. They'd ice 'em just the sairme as they do on the trawlers now, 'cept they dint put quite s' much on 'em. Soles an' plaice used t' be the things on board a smack. Soles an' plaice. Dorgfish you used t' chuck over the side quite orften 'cause they wun't worth much money. If you got plaice, an' they were in good condition, you could mairke some money on them. You dint git much stockie, though − only about a couple o' bob a trip. Mind yuh, if you mairde a hundred pound trip on a smack, that wuz very, very good. Very good indeed.

'You hed bunks f' sleepin' in on board. On the ow *Spray* the skipper an' the mate used t' spell each other round, but I dint turn out at all, not f' a watch. I wun't allowed t' take a watch, yuh see. Like that, I might git three or four hours in my bunk when we were towin' along, so that wun't too bad. Mind yuh, that wuz poor ow accomodation. Still, you used t' git a laugh or two. If you got a good bag o' fish, the ow man'd be right pleased. But if you got a poor one, oh dear − he'd hev a brill on f' the rest o' the day. Yeah, hev a brill on; ent very happy, we might say. We called that a brill. In the drifters the ow man'd git in the wheelhouse. If we wun't gittin' no herrin', he'd stand there in the wheelhouse an' scowl. I used t' look up an' I used t' say t' the boys, "Look at the ow man. He's got a brill on this mornin' all right, the ow bugger!"

'In a breeze o' wind them ow smacks used t' list right over. Thass what I allus said about 'em − the decks were practically scrubbed down on their own. See, the ow gal'd list over an' the water'd come through the scuttles, go acrorss the other side an' then come back agin. An' that allus kept the decks clean. When you were comin' in, you hetta stow everything up an' scrub round. You used t' use a broom f' that job an' some sort of soda stuff. What wuz that now? We hed that in the Navy durin' the war. I know − caustic soda! The ow man used t' put a little bit o' that down an' you used t' hetta scrub it round so that'd keep the decks white. Bugger gittin' down on yuh hands an' knees! I dint hev none o' that, though I know some of 'em did. Yis, some o' the ow skipers used t' hev the blokes, down, a-scrubbin' away.

'When you're a-coming' in an' there's a fair breeze, you keep all yuh sails up an' just sail in. They used t' open the bridge for yuh an' you'd go right through afore you stopped. Then you'd turn round an' git a tow back through the bridge. You'd use a drogue t' slow yuh down an' then a tug'd come an' tow yuh back through the bridge. When you got inta the dock, you'd heave yuh fish up out o' yuh hold on the gilson. You'd hev a little ow tackle on yuh foremast an' the fish'd come up in baskets an' git swung out onta the market. You'd land yuh fish one mornin' an' you'd go t' sea the next. You only hed one night in. I used t' go hoom t' Golstun on the bus. Yeah, An' I know one trip there, the ow man say t' me, "Come on. We'll go an' hev a drink." Well, I wuz never much of a beer drinker, but I went inta one o the pubs up there on the Low'stoft main street along o' him. That wuz the "Adelaide". I only hed about a pint an' my ow hid wuz a-goin' round an' round. I thought t' m'self, "I ent goin' t' hev no more o' that!"

'When I wuz on the smacks I got 10 bob a week an' a couple o' bob stockie. That wuz hard work an' all. I know you hed yuh steam, but there wuz a fair bit o' pullin' too. You dint hev winches like they do on the trawlers now; you hed a capstan. That hed wooden bits on the drum an' you used t' put yuh warp round that an' haul in. Thass all you hed an' that wuz driven by steam from orf the boiler. Yuh bread used t' swing in a basket up for'ad in what we called the fore-peak. That used t' swing about in there an' the last loaf used t' be as hard as iron. Mind yuh, you hed yuh sea biscuits an' all. Cooper's used t' mairke them. So did Sussams, but they were in Yarmouth. Cor, they were hard as iron, them ow biscuits. The Scotch ones were the best, lovely gret big round things, but ours were as hard as iron. An' full o' flour! You used t' hev 'em in a sack down in one o' the grub lockers an' they'd be full o' them little ow silver things. Weevils. Yeah, there'd be plenty o' them there. Bugs wuz another thing. There wuz plenty of bugs aboard the smacks. They were the big ow brown ones an' you used t' be full o' bites from them. When there got t' be too many aboard, they used t' fumigate the boat out t' git rid of 'em.

'I lay there in my bunk one night in the *Spray* an' I looked up an' I could see the moon. There wuz a gap betwin the planks. I told the ow man. I say, "You just look." He got in my bunk — "Coo t' hell, yis!" he say. "We'll hetta hev that done when we git in." We bunged it up wi' canvas there an' then. You know, knocked canvas inta the crack. An' when we come hoom, they done it prop'ly wi' pitch an' that. Yeah, I laid there an' I could see the bloody moon! Course, when you're in them smacks an' there's heavy seas, one minute you can see all the smacks around yuh an' next minute you wun't see 'em at all. You'd be down in the dip. An' then up yuh come again. They were good ow sea ships, though. Oh blast, yis! I mean, some of 'em done well in the First War. They hed guns aboard 'em an' they went as submarine chairsers!"

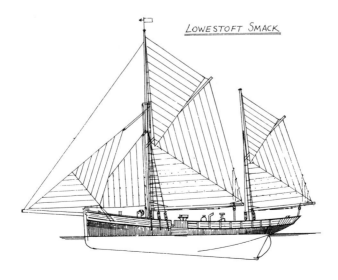

LOWESTOFT SMACK

The men in the photograph above never saw the result of their visit to the photographer in August 1917. Immediately afterwards they went to sea in the armed smack *Ethel & Millie* and met the terrible end described in the following chapter. Seated in the armchair is Skipper 'Johnsey' Manning. It has not been possible to identify his crew. **Right**: Skipper Tom Crisp, D.S.C., V.C. commander of the *Nelson*. **Left**: Tom Crisp, junior, mate of the *Nelson* at 16 (he told the Navy he was 18), outside Buckingham Palace after receiving his father's posthumous V.C. and his own D.S.M. from King George V. He wears his father's medals on his right breast.

CHAPTER THREE

Haul King George's Trawl

The Smacks at War

'If upon your port is seen
A little German submarine,
Do not fire shot or shell —
Just turn around and run like hell!'
(Fishermen's parody on a
navigation rhyme)

Throughout history the fishermen of Britain and their ships have formed a ready made naval reserve. The outbreak of the First World War found them at their peak with nearly 100,000 men employed in the industry. Under contingency plans made a few years before 200 steam trawlers were immediately requisitioned for minesweeping and patrol duties, the first of 3,000 fishing vessels eventually taken into naval service.

Roughly half the fishermen went with their ships into the Royal Navy, but the nation needed food and fishing had to continue as best it could. Those who stayed to fish lived no less dangerously than their mates in the Navy. The British fishing fleet suffered heavily at the hands of German warships and submarines and 'the price of fish', even in peace time an expensive one in terms of human lives, became even dearer. During the four year conflict 670 fishing vessels were sunk and 440 lives lost.

The trawlers and drifters requisitioned by the Navy were of course the latest and best steam vessels. They had no use for the hundreds of sailing trawlers then surviving, or so it seemed at first, but events proved differently. A handful of Lowestoft sailing smacks eventually fought and sunk a number of German submarines in one of the oddest and bravest episodes in the history of fishermen in wartime. One Lowestoft skipper earned a posthumous V.C.

So far as I can ascertain Lowestoft was the only port to have armed sailing smacks, though steam fishing vessels were fitted with guns in a number of other places. The sailing smacks were of course sitting ducks for submarine commanders who did not waste expensive torpedoes on them. When they located a smack, they surfaced near it, sent over an armed party in a dinghy, ordered the crew off into the little boat, took whatever stores and provisions they fancied, and then left a bomb on a time-fuse down in the fish-hold or chain-locker. The resulting explosion sent the smack to the bottom of the sea — by which time the U-boat had often submerged and gone. Some incidents were of a less merciful nature.

One should not equate the First World War submarines with the sleek, high-powered predators which terrorised the Atlantic convoys in World War Two. The early U-boats were slow, cumbersome affairs, capable of carrying only a couple of torpedoes and equipped with a small gun. It was this comparatively puny armament and slow speed which convinced at least one Lowestoft smack owner that any meeting between fishing smack and submarine needn't be completely one-sided. Skipper Fred Moxey had had enough of seeing his home-town's trawlers getting blown up, so he suggested to the naval authorities in Lowestoft that a few boats be equipped with guns in order that they might defend themselves in a manner which the enemy wouldn't forget. The proposition wasn't regarded seriously at first, but the continued sinking of fishing boats ultimately convinced the powers-that-be that there was nothing to lose by trying it.

To this end, four of the Lowestoft smacks (including two of Fred Moxey's own, the *G & E* and the *Telesia*) were fitted out with three-pounders and they proceeded to give a good account of themselves. In fact, the *G & E LT649* sank her first submarine only a couple of days after leaving port on her "maiden" trip as an armed vessel. Between official reports and faded local lore it is difficult now to assess the real degree of success of this minor campaign, but it was not negligible. It was also not without reprisals. In August 1917 the crew of the armed smack *Ethel & Millie LT200* (Skipper Johnsey Manning) were lined up on the fore-deck of a submarine, which then proceeded to dive. Sinking without warning became a feature of the U-boat war on the fishing fleet.

The men who remember first hand this side-show of the 1914-18 war at sea are now few indeed. One of them is Ted Fenn (born 1898), who was cook on the *Nelson* in which skipper Tom Crisp won his posthumous V.C. Ted Fenn is certainly the last survivor of this action which is as familiar to Lowestoft children as Trafalgar. This is his account of the loss of the *Nelson* and the *Ethel & Millie:*

'I wuz workin' on the land when war brooke out an' then I joined the minesweepers later on. I went in a Grimsby trawler, the *Daroogah GY191,* an' I suppose I wuz in her about a year, praps a bit longer. Then they starting askin' for volunteers t' go in these four armed smacks at Low'stoft. I said t' the skipper that I wun't mind goin' in one o' them 'cause they were givin' two shillins a day extra, danger money. Course, you were sorta pirates, yuh know! An' that wuz why there wuz that two bob extra. That wuz a right temptation, an' bein' so young an' silly I dint really know what I wuz doin'. I've often thought about it since, though! Anyway, I joined this smack, the *G & E,* about the 25th of January, 1917. She wuz layin' up at Chambers's yard, hevin' a better gun put on her − a 13-pounder, I think. That wuz on deck, but that wuz half-camouflaged so you couldn't see it from a distance. Once you were out at sea, you hed t' fish. You wun't really there after the fish, but you'd be usin' an ow beam trawl.

'Course, I dun't know if you know, but before they hed the armed smacks these German submarines were sinkin' the Low'stoft smacks hand over fist out on The Knoll, an' other places as well. They'd come up alongside an' order the crew inta the little boat, then git what fish they could, put a bomb down in the chain-locker an' blow the ship up. They dint waste a torpedo on 'em. No, they'd just come up alongside like I said. I mean, they were only fishin' boats; they couldn't do anything. Well, that went on for quite a while an' the smack owners, they began t' git a bit concerned. See, they were losin' so many boats. An' I think it wuz Fred Moxey, though I can't be sure, who went onta the South Pier t' see Commander Bruce an' ask him if some o' the boats could hev a gun put aboard so they could try an' protect themselves. Well, o' course the navy men made a laugh o' that at first. I mean, they were reg'lar service an' they thought it wuz no use puttin' a gun aboard a little ow sailin' smack. But anyway, the sinkins went on an' they sent f' Moxey after a while an' asked him if he still wanted t' go ahid with his idea. He said

Above: the hazards of war. A mine brought aboard by a steam trawler. It has been lashed to the fore gallus on the port side. Behind it is the large dahn buoy used to mark out a 'piece of ground' to be trawled.

Below: a First World War armed trawler leaving Lowestoft on minesweeping duties. The look out platform on top of the wheelhouse (a naval addition) was popularly known as Monkey Island.

yes, so this smack what I went on, the *G & E,* hed it done – an' so did about three more.

'Now this 13-pounder gun they were puttin' on her when I joined wuz a new thing out from what I understood about it. I think she'd hed a five-pounder or a six-pounder afore that. Course, the steam drifters used t' hev either three-pounders or six-pounders on the foc'sle hid, an' some o' the big trawlers used t' hev a 12-pounder.

'Now this new gun on the *G & E,* the muzzle wun't above four foot long. I dun't know much about gunnery, but they explained t' me that that wuz t' stop the vibration, yuh see, t' cut down the recoil. That wuz bad enough as it wuz. When we went t' sea t' test the gun, I wuz down below an' that felt as if we'd struck a blinkin' mine! The soot come down the blinkin' engine-room funnel an' everything jumped up. But, anyway, she stood it well enough t' do what we wanted. I dun't know if they reinforced the deck or anything, but I know the little boat helped camouflage the gun. If I remember right, the little boat wuz put alongside the gun so you couldn't see it from a distance. But, o' course, when you started t' use the gun, you hetta open everything out. When you were fishin', you hetta haul as best as you could. Specially at night, 'cause you wun't allowed a light. The idea wuz t' coy the submarines up to yuh. Well, that wuz all right the fore part o' the time, but after one or two submarines hed bin sunk Jerry got wise over that – an' he wun't comin' up alongside no more. He wuz a-goin' t' try an' catch yuh when you were too far orf.

'We actually went t' sea in the *G & E* the end o' January, 1917, an' we sank a submarine the first day o' February. I shall never f'git our position – 17 miles E.S.E. o' Low'stoft. I wuz down below, so I dint see a lot, but the crew said the skipper, Tom Crisp, wuz manouevrin' us about on the motor. See, we'd got a little petrol engine an' they started that up. We were with the *Telesia LT1155,* though her name hed bin changed t' *Boy Alfred* at that time. Well, so hed ours. We were known as the *I'll Try* 'cause the boat hed already sunk a submarine afore I joined her. They used t' change the names reg'lar, yuh know t' confuse the enemy. Apparently this submarine come up an' give a burst o' machine gun fire at the *Telesia* an' you could see where the bullets hed hit all along the trawl beam. Then he went down agin. When he come up the next time, the crew said he wuz comin' straight for us. They see him come straight up out o' the water an' they reckon the shell from our gun hit the connin' tower. An' o' course there wuz oil an' that all over the water, so that wuz a fair hit.

'Anyhow, after we'd done that, we put the dan overboard t' show where the submarine wuz sunk. Then the ow man, the skipper, he say, "We're goin' t' Southwold t' bring up under the land f' the night." On the way there we met some P-boats, fast submarine-chasin' jobs somethin' like a frigate. One of 'em headed right for us an' I can picture her right now. Cor, she wuz a-steamin'! You know, frothin' at the bows. An' when she got near enough, the captain or whatever he wuz spoke from the bridge through this megaphone. "Where's the submarine, skipper?" he say. Tom Crisp say, "Oh, we've sunk him." "Hev yuh!" say the officer. Then he say t' the matelots, the sailors what were there, "give three cheers," he say, "for the ow Low'stoft smack!" I can remember that as well as can be. An' o' course I felt as big as two people.

'Well, when we got t' Southwold, we brought up f' the night an' I can remember about a foot o' snow on the ground when we got inta Low'stoft the next day. We come in on the Sunday afternoon an' we went an' laid in the yacht basin. Then we hetta all go up onta the

South Pier an' say what we knew about the submarine business. Well, I dint know much about it really 'cause I wuz down below, but I knew that there wuz a thousand pound reward f' sinkin' a submarine, so I thought t' m'self, "Cor, I shall hev about 90 pound!" Yeah, I thought my share'd be about 90 pound an' I thought the skipper'd git about 120 or 130. There wuz nine of us in the crew altogether, where a smack wun't only built f' five. I wuz the cook, an' apart from me there wuz the skipper, the mate, the third hand, the deckie, three gunners an' the chap what looked after the motor. The gunners were naval gunners; they come orf the *Halcyon*. They were volunteers an' they'd done the same as me – they'd joined f' that extra two bob danger money, which they got on top o' their 2/9d a day or whatever it wuz. That extra two shillins went right through the ship.

'Now there used t' be an ow boy called Lomas, who used t' watch the smacks in them days, an' when we brought up in the harbour he come aboard an say, "You'll git yuh submarine money t'morrer." I dun't know how he knew, but we all thought, "Well, thass all right." Anyway, I got my duffle trousers on an' went ashore thinkin' I wuz in the money. Home I went t' Lound an' I told my father all about it. I say, "We're gorn t' git the submarine money t'morrer." He say, "How much do yuh think you'll git?" So I say, "Well, accordin' t' what I've heard, my share should be about 90 pound." He say, "Well, we'd better go an' hev a drink." So we did.

'When I went down t' Low'stoft the next mornin' an' jumped aboard the ow smack, all the blokes were arguin' an' goin' ahid. They were in the cabin. Well, I wuz cook an' I wun't allowed t' say right a lot, but I stood there in the engine-room, which wuz next door t' the cabin, an' I say, "Whass the matter? Whass all this?" One o' the older chaps say t' me, "We aren't a-gorn t' git our submarine money now!" So I say, "Why not?" He say, "Well, our skipper an' the *Telesia's*, they're both bin up on the pier, an' the *Telesia* is goin' t' git the thousand pound. We're a-goin' t' git 200 pound an' an' award f' gallantry." O' course, these naval chaps were really put out an', I mean, we were the one that sank the submarine. There's no doubt about that. Anyhow, we hetta go up Bank Chambers at such an' such a time an' I finished up gittin' 17 pound – instead o' 90!

'Well, after that business they changed out name from the *I'll Try* t' the *Nelson* an' we went t' work agin, yuh see, fishin' merrily on until the summer frap. well, o' course, we got sunk that time. We met our Waterloo. He wuz too much for us. We couldn't do nothin'; he had a four inch gun. Or wuz it a 4.7? That wuz a lot bigger'n ours anyway. Course, in betwin the February an' the August we hed a nice little run. We never saw a submarine durin' that time, so o' course that gave me confidence. I thought that wuz money for jam. An' another thing – the submarine we sunk in the February, he only hed a little gun, but this one in the August, oh dear, he wuz a different proposition.

'Now I can't give yuh the full details o' what happened, but we got sunk on the 15th o'August, 1917. That wuz the afternoon an' I wuz cleanin' the fish f' the next mornin's breakfast. I'd done the washin' up an' squared up after dinner, an' I wuz cleanin' these fish when Tom Crisp, the skipper, come up an' hed a look round. He say, "Fetch me the glasses." So I went an' got 'em orf the mizzenm'st an' he hed a look through 'em. He say, "There's a couple o' submarines out there," I think. An' he kept lookin' f' quite a long while. Then he say, "Yes, give 'em all the down. Tell 'em t' be prepared f' action." Well, my job in action stations wuz down below in the fore-peak, where the ammunition wuz.

We hed 50 rounds o' these 13-pounder shells in boxes. I dun't know whether there wuz four or six in a box. That wuz my job t' pull them out an' hook 'em inta two hook-ropes what'd bin passed down from on deck. When I'd done that, the blokes'd pull 'em up.

'Well, I wuz a-doin' that when this U-boat opened fire. An' I can't remember exac'ly, but I think that wuz about the fourth shell what hit us on the starboard bow just above the water-line. An' if that'd come just a little further for'ad − well, that'd bin me gone. I can still remember the water a-splashin' an' I can see the flash as well. Then that went quiet f' a bit. Our gunners knew he wuz out o' range an' the skipper told 'em not t' fire an' t' hang on f' a while. Praps they might come a little nearer. but o' course they dint. They started firin' agin. There wun't many shells come over, but one went through the mains'l, an' we fired about four or five rounds in reply. Well, our gunner said afterwards we fired about four or five. The next thing I remember is one o' the blokes sayin', "They're killed the skipper!" O' course, we were fightin' a losin' battle. This shell hit the skipper direct. He wuz standin' at the tiller when that come over an' that nearly cut him in half. I can picture him layin' there now. That hit him an' went right through the port quarter.

'Well, there we were − the ship a-sinkin' an' the skipper done for. He wuz still conscious, though, an' he said t' his son, young Tom, who wuz mate, "Abandon ship. Throw the books overboard an' throw me over after 'em." Course, you hed these confidential books an' papers, or the skipper did, an' he dint want 'em t' fall inta enemy hands. An' the blokes told me afterwards that that wuz what he said, "Throw the books overboard an' throw me after 'em." Well, o' course young Tom wanted t' git him inta the little boat, but he wun't let him, so we left him on board in the end.

'We launched the little boat when I'd come up from the ammo room. Then I went down the cabin. Somebody told me − I f' git who it wuz − t' git the tea kittle full o' water. I went down inta the cabin an' the water wuz about up t' my knees. I just managed t' git the kettle under the tap an' I filled it up an' took it back on deck. Then Rosso, the gun-layer, he got one o' the pigeons out o' the little boat. See, we hed this coop o' pigeons f' messages. Him an' me went back down the cabin an' on the table wuz this special paper. I remember him writin' on there. "Armed smack 'Nelson' attacked by submarine − Jim Howes Shoal Buoy." Thass near the Leman Bank, about 15 mile north o' Smith's Knoll. Thass where we'd bin fishin'.

'I actually held the pigeon time he writ the message out. Then he stuffed the paper through the ring on its leg an' let it go. Well, o' course, that blinkin' bird dint go straight f' hoom! That went an' lit on some smacks what were fishin' at The Knoll. Anyhow, they scared it orf an' that did eventually come t' Bagshaw's lorfts (I think it wuz Bagshaw's). Anyway, they got the pigeon an' the message went round Low'stoft, so they sent out two boats t' look for us. They were somethin' betwin a light cruiser an' a destroyer an' they were stationed at Low'stoft. There wuz the *Halcyon* an' the *Dryad,* an' they sent them out t' look for us. That wuz about three o' clock on the Wednesday afternoon when we got in the little boat an' pulled clear. An', o' cqurse, Jerry wun't very partic'lar. There wuz a few shells droppin' around after we'd got orf. Well, he'd fired on us first of all 'cause he knew we were an armed smack. Yes, he knew what we were all right!

'Then, as luck would have it, that come over a thick fog. One o' them sea mists you git in the summer. That wuz really a miracle, wun't it? We were a-dodgin' around an' we could see this submarine, but he couldn't see us. Our partner wuz the *Ethel & Millie* at this time. When we sunk that submarine in February, they commandeered four more smacks an' hed a reshuffle. So instead o' us bein' with the *Boy Alfred,* we were with the *Ethel & Millie.* Well, there we were in the little boat an' Jerry knew we couldn't git far, so he concentrated on the *Ethel & Millie.* The last we see of 'em, he wuz moored up alongside her an' we all thought he wuz a-gittin' some fish out. The crew, Johnsey Manning an' them, were lined up on the submarine's foredeck. Well, then that come over really thick, so we never see n' more o' the *Ethel & Millie.* An' yit dint anybody else! No doubt Jerry destroyed her somehow. An' o' course the worst thing about it wuz we never heard no more about the crew. I've got a photograph o' them hangin' up in my garage now.

'Like I say, that wuz three o' clock on the Wednesday afternoon when we got in the small boat, an' we were dodgin' around till next mornin'. Then we thought we'd start t' make for Low'stoft. We'd got the ow compass from orf the *Nelson,* lifted it out an' put it in the little boat, but the action o' the little boat wuz too quick for it. Too much jerkin' about. But we did hev the North Star, an' we reckoned that if we kept that on the starboard bow we'd be headin' westward an' comin' somewhere towards land. Each of us took turns on the oars, an' the ow boat wuz leakin' like a sieve 'cause she wuz only built f' five an' there wuz eight of us in her! So there wuz one of us a-balin' the water out an' two a-pullin' on the oars. The rest laid down an' tried t' git a little bit o' sleep. We were like that two days an' two nights. Then the ow *Dryad* picked us up, one o' the boats I wuz tellin' yuh about. They'd got the message, I spose, an' come out looking for us.

'Afore they come along, we sighted this buoy. Well, we thought that wuz one o' the buoys layin' by the Haisbro Sand, but when we got near enough we could see it wuz the Jim Howes Shoal Buoy! We'd bin pullin' two days an' two nights an' we were still in the same place! The buoy wuz a cage buoy an' that used t' hev a bell in it, but they'd took that out durin' the war an' took the name orf. We knew it, though, an' any ow fisherman'll tell yuh where it is. Anyhow, some of us wanted t' keep pullin' an' some wanted t' make fast t' the buoy. In the end we pulled up to it an' decided t' git the painter an' tie up to it. That way we could at least git some rest from rowin'. Well, young Tom Crisp, our mate, (later awarded the D.S.M.) he jumped aboard this ow buoy. You can just imagine how big they are, but he managed t' git on it. The only trouble wuz that the painter dropped in the water. There wuz a strong flood tide a-runnin' an' we started movin' away like anything. We pulled an' we shoved an' we just managed t' sling this rope. Tom got hold of it an' tied it round the buoy an' thass where we hung. Tom wuz on the buoy an' us other seven were in the little boat. She hent got much more'n about a foot freeboard an she wuz leakin' an' all.

'We hed sin some minesweepers on about the Thursday an' we tried t' attract their attention, but that wun't n' good. We made fast t' this buoy about noon on the Friday an' there wuz a real tide runnin'. Thass a marvellous thing that dint pull the stem out o' that little boat. I very orften think o' that. You know, seven men in her an' that tide a-runnin'. That coulda pulled the stem out or brooke the painter. But, anyway, after a bit we all see this smoke an' all see this ship, an' as she come nearer we could see that wuz a warship. So Rosso, the gun-layer, he could semaphore an' he got up in the boat an' semaphored, ''Crew o' the *Nelson*''. Well, this ship come steamin' up an' young Tom Crisp managed t' jump aboard her. He jumped inta the bight o' the anchor chain where that come out o' the hawse-pipe. The anchor wuz up on deck, yuh see. Yeah, he wuz the first one what got aboard. Well, o' course, when they got further along, they put the steps down so the rest of us could git up.

'Well, when we got on board, the sailors reckoned that wuz a lucky thing f' us that they dint put a shell into us! I mean, when you come t' think about it − a buoy settin' there an' then this boat hangin' from it. That looked like a submarine t' them at first! They thought that wuz the submarine what'd done the damage an' they were very near puttin' a shell into us. Well, if they hed a-done, that'd bin it! But, anyhow, they brought us inta Yarmouth Roads, anchored there an' signalled ashore for a boat. They sent out the *Chris SN118,* a North Shields drifter. She wuz one o' the patrol boats stationed at Yarmouth an' she brought us inta Low'stoft. I hent even got n' boots on. I'd slung them orf when we abandoned ship. When we got in, the word hed got round that we'd bin sunk an' people knew that we were in great difficulties. The only lucky thing wuz that we wun't due in till Friday anyway, so we wun't over time. The butcher what used t' go out inta the country an' supply us wi' meat, Cullen, he hed little ow toshers as well, so he knew all about it when he went round that week. But, o' course, he dint say nothin' t' my fam'ly.

'After we got sunk we all got about nine or ten days leave. Survivors' leave, they called it. Then we hetta report back t' work. We hetta report down at Phillips's Stores. That wuz a place up-through-bridge. You got at it from Commercial Road, near the dry dock, an' thass where the minesweepers an' the patrol boats got their provisions from an' all that sort o' thing. Well, the blokes there used t' make it nice for us. The butcher, he wuz a marine, an' he used t' see we got good meat, so we used t' see he wuz all right f' fish. Then we used t' come in f' all the seasonal stuff, fruit an' vegetables an' that sort o' thing, so we used t' live pretty well. Then, when I went down there one mornin', they said we'd all gotta report on the South Pier an' see draftin' officer West. He wuz a fleet reserve man an' he used t' do all the draftin', findin' crews an' that. Well, he told me that I'd gotta go down t' Chatham Barracks an' that all the rest o' the crew hed gotta go t' Kirkwall, in the Orkneys. An' o' course I wuz really disappointed about that, but that wuz what happened an' I finished up at Malta in the end. I wuz there right till after the war finished.

'Now on the armed smacks you wun't allowed t' go near the boat wi' uniform on. That'd give the game away. So I went t' Long, the fishermen's outfitter an' got two pair o' duffle trousers (all the fishermen used t' wear them), a jersey an' a wrapper. I thought I wuz ever so clever. Then I went t' Saunders, who used t' make the fishermen's boots, the leather ones. I'm not talkin' about the long sea-boots, but the little ones with the high heel an' the fancy toe-cap − you know, like with the pattern of a heart or an anchor. Yeah, I got a pair o' them for best. Well, when we went away on that last trip, I left my ordinary boots, my navy boots what'd bin dished out from stores, I left them behind t' hev suffin done to 'em. So the day we got sunk I hed these bran' new boots on an' they were lorst wi' the ship!

'The owners o' the armed smacks got charter money from the navy, but they dint git the money from the fish what were caught. We were fishin' f' the gover'ment an' that fish wuz sold on the market an' we dint see the money for it. Mind yuh, we used t' git a good bit o' stockie. That used t' work out about ten bob a week wi' the gurnards an' all the small stuff we were allowed. We used t' be at sea about six days an' our actual wage wuz 3/6d a day. Thass what all deckhands on minesweepers an' patrol boats got. But we got this extra two shillins danger money as well, so I wuz gittin' 5/6d a day. Well, that wuz the enticement. Yis, that wuz what I wuz after. I wuz thinkin' more about that money than I wuz about anything else. but I dint really realise what I wuz goin' on, bein' young an' inexperienced. Our lot used t' be known as The Trawler Section. We hed a badge with RNR on it an' underneath a 'T' for trawler. They used t' call us Harry Tate's Navy. These fleet reserve chaps we hed with us from orf the *Halcyon,* they only got 2/9d a day, plus the two bob danger money, an that wuz a bit of a sore point with 'em. See, I wuz better orf'n them an' I wuz only a kid. Mind yuh, they used t' git the stockie bait. That used t' be shared out all round the crew.

'Like I said, you'd be out for five or six days an' then in for two or three. We dint see that many mines, yuh know. You'd see the odd one afloat, but not many. One o' the biggest problems wuz hevin' nine people on board. There wuz only five bunks in an ow smack — two on each side o' the cabin an one acrorss the stern. Well, o' course when that come to it, there'd be one or two of yuh on watch, an' one or two doin suffin else, so there wuz enough bunks f' everybody really. Once you'd shot yuh trawl, you used t' lash the tiller an' tow along. Then, when the skipper thought you'd towed long enough, he'd come an' call yuh out. "Haul King Gerorge's trawl!" he'd holler, an' you'd all turn out. Course, if that wuz dark, you wun't allowed t' show a light, yuh know. But you got over it somehow. My job wuz down below a-coiling' the trawl warp. Coilin' the trawl warp, yeah. That wuz a rum job too. The cold water'd be a-runnin' down yuh arms! I never hed anything t' do wi' shootin' the trawl, though. I'd be turned in then. I dint hev anything t' do wi' that.

'One good thing on board wuz the grub we used t' git. Like I told yuh, bein' friends wi' the people at Phillips's Stores saw us all right. An' we used t' git jolly good fish breakfasts on board as well. Then there wuz the mess savings we used t' git. We used t' be allowed so much f' victuallin', an' if you dint spend all your provision allowance you used t' git the difference back as mess savings. An', o' course, bein' well in at the stores, we could allus keep under our allowance. Sometimes you might make about half-a-crown or three bob a trip, so we were in clover all ways. Oh yeah, I used t' do very well when I wuz in the *Nelson.* Well, like I told yuh, I joined in the first place f' that extra two bob a day. The only thing wuz I dint really see the danger. That wuz all nice an' easy till we got sunk.'

Grimsby Smack C.1880
70 tons.net

Above: the motor smack *Pathway LT397,* 28 tons, after her conversion from sail. She was fitted with an 80 h.p. diesel.
Below: the steam drifter trawler *Grey Sea LT1279,* said to be the first Lowestoft fishing vessel to be fitted with a radio receiving set in the early 1920's.

CHAPTER FOUR

Life In The Old Girl Yet

The Motor Smacks

'When the sun goes down as clear as a bell,
An easterly wind, as sure as hell!
But when the sun goes down behind the black,
A westerly wind you may expect.'
(Traditional − East Anglian fishermen's
weather rhyme)

The 1920's and 30's were generally bad years for the whole British fishing industry, compared with the expansion of its earlier decades, but one significant change that occurred was the adoption of diesel engines. Increasing coal costs had a good deal to do with it, as well as the inconvenience caused by miners' strikes in 1921 and 1926.

The earliest use made of either petrol or hot-bulb oil engines was mainly by Scottish sailing drifters before 1914 but it was several years before boatowners in other parts of the British Isles went in for marine internal combustion engines on any scale.

One of the first ports to try them was Lowestoft, perhaps given an impetus by the fitting of auxiliary engines to the armed smacks. The *Veracity LT311,* Britain's first purpose-built diesel herring drifter, was constructed in the local Richards yard during 1926 and this was followed by a noticeable switch to diesel engines in the trawler fleet. This involved both the conversion of a number of sailing smacks to diesel power and the construction (again by the Richards yard) of a small fleet of specialist, steel-hulled, near-water trawlers driven by Ruston-Hornsby engines. These were the 'Ala' class of boats which became well known in local lore. We shall not deal with those here but with the converted smacks, for the diesel power gave them a new lease of life.

According to 'Olsen's Fishermen's Almanac' for the year 1937, there were 18 of these so-called motor smacks registered in the port, as opposed to 60 of their sailing counterparts. A dozen of the 18 were owned by three associated firms, W. H. Podd Ltd., Diesel Trawlers Ltd. and Inshore Trawlers Ltd. For the most part, the boats converted were the most recently built smacks (1920-22) because their timbers were best able to stand the stresses of having an engine-room thrust into their bowels and a propeller shaft running out through the stern post. There were a few that were older, most notably the *Dorando Pietri LT295,* which was named in honour of the Italian long-distance runner, who failed so nobly in the marathon in the 1908 Olympic Games.

To begin with, the diesel engines fitted were small ones of around 75 HP or so, manufactured by two firms − Deutz and Allen. For the most part, these were auxiliary engines, a supplement to the sails, but later conversions used larger power-units of 100 or 135 HP and the mains'l was done away with. This second kind of conversion produced the true motor smack, and as well as the Deutz and Allen engines, the names of Crossley and Petter also began to be heard. In fact, the conversion that the Richards yard in Lowestoft did on the second *Purple Heather LT249* in 1934 included the first marine diesel made by

Crossley. Most of the sailing smacks converted to motor-power were in the 40-46 net tonnage range, and the provision of an engine-room below decks brought this down to 27-36 tons.

The fishing range of these boats was not great and the small engines necessitated the use of light gear. Working out of their home port, they would travel as far as the Gabbards, in one direction, and the Leman Bank/Hammond's Knoll area in the other. Apart from this, some of them used to go on the traditional spring voyage for soles round to Padstow, because once the long journey to the Westward was over, there was good fishing to be had just off shore. This particular voyage became quite important during World War Two and continued for a year or two afterwards. In fact, a handful of motor smacks carried on fishing out of Lowestoft into the 1950s. One of the last was the *Pilot Jack LT1212*. Jack Rose (born 1926) remembers working in her:

'On the ow motor smacks you used t' go through the hatchway, down a flight o' steps an' straight inta the galley. For'ad o' the galley wuz the engine room an' aftside wuz the cabin. The engine wuz actually in the aft side o' the hold. I dun't know what make o' engine we had in the *Pilot Jack,* but that wuz a diesel. In the cabin itself you were right cramped up. That wuz very small, a lot different t' the drifters an' the trawlers what I wuz in. See, bein' a converted smack, that wuz small t' start with. Underneath the cabin table you had what they called a jenny − a generator. That wuz for the lights an' that'd be goin' all trip. That used t' spit out diesel fumes all the time an' when you turned out o' yuh bunk, yuh mouth used t' be all coated wi' diesel oil! Oh, that wuz terrible. As you lay there in yuh bunk, that'd be belchin' these fumes out all the time. But people used t' stick it because they like bein' on the little ow boats.

'I went in the *Pilot Jack* after I come orf the steam trawlers an' drifters. That'd be back in the early fifties. We used t' go down in the mornin', git the ow boat riddy an' the gear riddy, an' then we'd set out. We used t' go down t' Southwold sometimes. An' if we dint go south-about, we'd go north-about. You know, down t' the Haisbro or the Cromer Knoll an' what they call The Shoals. We used t' fish a lot in that area. An' as soon as that come on a scuffle − you know, a bit o' wind an' a bit o' swell − an' the ow skipper thought that wuz gorn t' git bad, hoom we used t' run. We had a wheelhouse on her, but that wuz only a small affair. You could only git two of yuh in there. Course, there wun't no sail on the forem'st. That wuz just there f' the gilson t' haul the cod end up an' do the natural sort o' work same as an ordinary trawler. In fact, they were just like a small trawler.

'I wuz deckie-second on her, an' besides me there wuz the skipper, the mate, the third hand, the chief engineer an' the cook. So there wuz six in the crew. We were a happy-go-lucky bunch. Yeah, I spose I wuz on about three or four quid a week wage. Then we all got 3d in the pound poundage. That wuz 3d in the pound on the clear hundred, after you'd paid expenses. Then we got ten bob stockie on the clear hundred on top o' that. The skipper an' the mate were on a share, but the rest o' the crew got a wage, plus poundage an' stockie. We used t git all the little ow rooker f' stockie, gurnards an' that sort o' thing. Afore the war that used t' be called stocker bait, but then that got cut short t' stockie.

'My uncle, Arco Rose, wuz skipper, but that dint count f' nothin'. "We aren't relations when we're at sea," he used t' say. An' that wuz proved when I hed a boil come up under

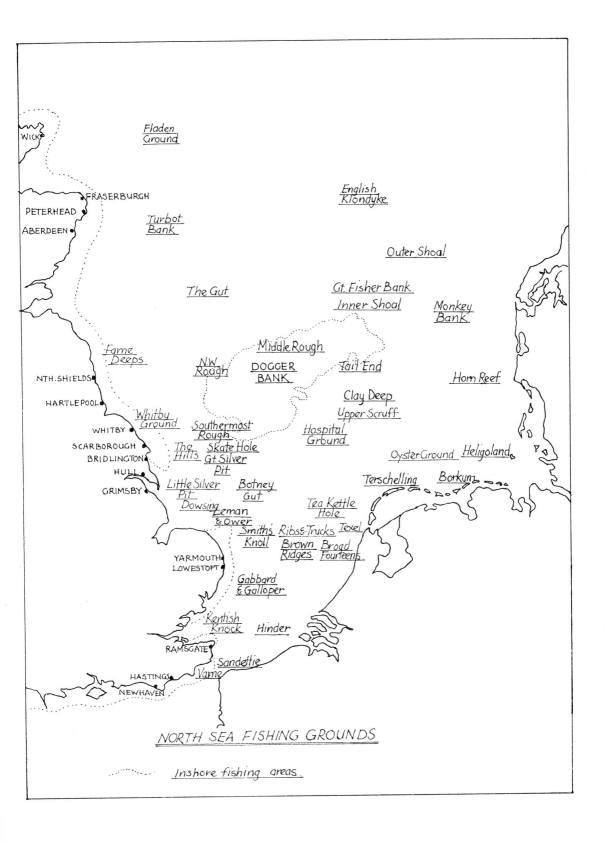

WICK

FRASERBURGH

PETERHEAD

ABERDEEN

NTH.SHIELDS

HARTLEPOOL

WHITBY

SCARBOROUGH

BRIDLINGTON

HULL

GRIMSBY

YARMOUTH

LOWESTOFT

RAMSGATE

HASTINGS

NEWHAVEN

Fladen Ground

English Klondyke

Turbot Bank

Outer Shoal

The Gut

Gt. Fisher Bank

Inner Shoal

Monkey Bank

Middle Rough

Farne Deeps

N.W. Rough

DOGGER BANK

Tail End

Horn Reef

Clay Deep

Upper Scruff

Whitby Ground

Southermost Rough

Hospital Ground

The Hills

Skate Hole

Gt. Silver Pit

Oyster Ground

Heligoland

Little Silver Pit

Botney Gut

Terschelling

Borkum

Dowsing

Leman & Ower

Tea Kettle Hole

Smiths Knoll

Ribs & Trucks

Texel

Brown Ridges

Broad Fourteens

Gabbard & Galloper

Kentish Knock

Hinder

Sandettie

Varne

NORTH SEA FISHING GROUNDS

.... *Inshore fishing areas.*

my right eye. That got so bad that ow Woggy Blowers, the mate, lanced it one day when we were in the wheelhouse. "I'll cure that for yuh, boy," he say. "Look out o' the winder." So I did. He got his shut-knife out o' his pocket an' he just slit acrorss my cheek. Well, the pus flew everywhere! But that eased the pain. Two days later my eye wuz black an' blue. I could only sleep on one side. We got called out that night t' haul an' I slept on. The crew left me down there. Next thing I knew, the ow skipper wuz a-shakin' me all ways. "Aren't yuh gorn t' do no bloody work t'night?" I swore at him, but I went up on deck an' stuck the rest o' the trip out. When we got in, I hetta go up the hospital with it.

'We used t' go f' weekly trips on her, an' sometimes you wun't even out that long. If the weather wuz bad, that'd be day b' day. I know we once went out in bad weather an' she iced up. They took photographs of her when we come in. You couldn't even see the block in the fore-gallus; that wuz just one solid lump o' ice. An' the ropes, they looked like posts. When we went t' shoot, we couldn't manage it 'cause all the net wuz frozen t' the deck. Oh, that wuz a bad winter, that year. You hear people say salt water dun't freeze. Thass a load o' bull, because it do. You ask any o' the blokes what went up t' Iceland!

'We used otter gear on the *Pilot Jack,* but that wuz on a smaller scale t' the big trawlers. You still worked yuh bobbins an' yuh titlers; but where the drifter-trawlers hed three or four titlers, we only hed the one. That used t' go acrorss the foot o' the trawl an' that used t' riddle in the ground. That used t' riddle the soles out. The net wun't all that big an' I know we used t' work a trawl each side. Yeah, an' we all used t' hull in an' git mended. We used t' fix our own gear up as well an' everyone'd turn out t' haul. Even the ow cook used t' come on deck an' haul. Yeah, poor ow Olley. He wuz bloody nigh 70 an' all, if not over!

'Course, if you had a good cook, you had a good crew. An' we did hev a good cook, although the ow boy wuz a bit orf it in the end. He wuz so old, yuh see. One day he upset the dinner on the coconut mattin'. The boat rolled an he shot the dinner everywhere. Well, he just got down an' scooped it up orf the coconut mattin' an' put it back on the plates. When we got it, there wuz all these bloody whiskers stickin' out of it! Ow Arco come an' slung it overboard. Yeah, he chucked the bloody lot over the side! I shall never forget that. The ow boy just got down on his hands an' knees an' scooped it up in his bloody hands an' bunged it back on the plates. An' there wuz all these whiskers stickin' out!

'Like I told yuh, I wuz deckie-second on her, so the ow chief used t' tell me what t' do regardin' the engine. He'd bin in that boat donkey's years, so we never had no trouble with the motor. Water wuz the main trouble. She used t' leak all the time. When my watch started, I used t' go straight down inta the engine-room, straddle-leg the dill an' start workin' the hand-pump. That wuz a bloody gret thing an' I used t' hefta give a thousand pumps on that t' clear her. That wuz enough t' clear her, but o' course she wuz a-fillin' up all the time. Mind yuh, I dun't think the chief used t' pump her on his watch. I got the bloody lot! Yeah, I weighed it up that he used t' leave it on his watch f' me t' do on mine.

'Another thing I hetta do on my watch wuz nip up on deck now an' agin t' keep an eye on the warps an' the towin' block. Yeah, I used t' watch the engine, then nip up on deck an' watch the trawl. The watches were about three or four hours long. We used t' work it among all the crew − except the cook, that is. He dint hev a watch. Sometimes you'd feel the warps come tight. The whole ship would shudder an' the engine would race. You knew you were on a fastener then, so you knocked the warps out o' the towin' block as soon as possible. The ow man'd stop everything. Then, soon as you'd knocked the warps out o' the block, he'd try an' bring the boat round t' clear the wreck or whatever you were hung up on. Sometimes there'd be s' much tide an' swell he just couldn't bring her round against the tide. That used t' be awkward then. When you were on watch an' wanted t' nip down below t' look at the engine, you used t' lash the wheel so you kept t' the course you were on.

'We used t' hev a trawl basket up the forem'st when we were fishin'. That wuz f' durin' the day. You used t' hev yuh lights at night. You hed two f' fishin' with. There wuz the triplex light, which hed a white centre an the red an' green either side. Then there wuz the all-round light, which wuz a white one that could be seen all round the horizon. We come inta harbour wi' that on once. We shouldn't ha' done. We shoulda hed the mast-hid steamin' light on, but we didn't — we hed this all-round one a-showin'. As soon as we berthed, we all got a rocketin' from the customs bloke. Yeah, because we were breakin' the navigation laws.

'You used t' tow along about three or four hours. That all depended on the skipper. If you wun't gittin' a lot o' fish, you'd hev a longer tow. If you were gittin' a good bit, you'd hev a shorter one. You always went with the tide o' course, but you never got very big hauls. No, no, nothin' big at all. If you come in' with 150 kit after a full trip, that wuz good in her. Sometimes you'd only git 70 or 80. See, you couldn't go where the big boats went; you had t' go south-about or north-about. An' you always kept in sight o' land. Yis, that you did! Oh, we could always see the land in the distance. Well, the ow skipper hed his various landmarks t' go by an' he wun't lose sight o' them if he could help it.

'We used t' run up t' the Gabbards in the ow gal sometimes. You used t' git a lot o' that peat stuff there. Moorlog, they used t' call it. I always used t' break it open on the deck. There used t' be a helluva mess, but that wuz interestin' t' look at. You used t' git coal an' all; an' if we trawled up any coal, we used t' stack it on the foredeck an' then we'd share it out when we come ashore. Oh yeah, you used t' trawl up a lot o' coal what'd come orf the various ships. Some o' the fishin' boats used t' go out o' Low'stoft with deck cargoes o' coal. Well, when they got out, the wind would sometimes git up an' that'd sweep the decks an' they'd lose all their coal! Then there were the ships what sunk. They all carried coal. Oh, you used t' trawl up no end o' coal. A lot of it used t' be encrusted wi' barnacles, but thass terrific stuff t' burn. You dun't wanta buy any coal when you can git salt-coal. No, that you dun't! We used t' put it in baskets, what we got. Yis, there's bloody coal all over the North Sea.

'North-about used t' be one o' ow Arco's favourite trips. Yeah, we used t' run down t' the Cromer Knoll. Thass where his grounds were. He wun't go no further. I spose he coulda done, but that wuz his favourite place. We used t' work the Leman Bank an' all. An' The Knoll. Yeah, we used t' work what they call The Knoll Gateway. Thass betwin the Knoll Lightship an' the Knoll Buoy. We used t' git plaice in there, but they used t' be thin plaice. Very, very thin at times. But, there you are, you hed t' git in where yuh could an' grab a livin'. We even used t' work just out orf Low'stoft on the Barnard certain times o' the year. You used t' git a little run o' cod there sometimes durin' the winter.

THE TRAWLERMEN

'She wuz a good ow sea boat, the *Pilot Jack*. The only thing wrong about her wuz that she did tend t' leak. That mighta bin the reason why the ow man run f' port every time that scuffled! Yeah, he used t' race f' port every time that come on scuffly. Eileen'll tell yuh that. I reckon I used t' hev more time in port than I did out! The owners dint mind, though, because you were out agin the next mornin'. If you stopped out in her in bad weather, you couldn't fish, so the ow man'd say, "Bugger layin' out here all night! We'll hev a night in our beds." Then early next mornin' we'd be down an' out agin, see. The owners, Podd, used t' accept it. They were a good firm t' be with. Yeah, they were people you dint mind sailin' for. They were a lovely little firm, an' o' course they used t' do all their own engineering as well.

'The gear on the *Pilot Jack* used t' be dead easy t' work. That wuz light, yuh see. You only used t' work the little ow wooden bobbins on yuh ground-rope an' there'd be a cowhide on the cod end t' stop it chafin'. Apart from guttin' up, there wun't much t' do as yuh went along – just makin' up a lazy deckie an' splicin' a quarter-rope an' that sort o' thing. Haulin' used t' be fairly easy simply 'cause the gear wuz light. You used t' heave yuh cod end up on yuh gilson just the same as on the bigger boats. The winch wuz only a littlun an' that wuz driven orf the motor by a belt. There wuz a belt come orf the front o' the engine an' that went up throught the ceilin' t' the winch. The chief used t' see after that when it got slack. He used t' tighten it. He used t' put these steel clips in, like teeth, an' draw it up. Then he'd hammer the teeth down flat. That wuz a good little ow winch. That wuz never any trouble. The ow chief, he used t' go round an' oil it regular. He used t' hev a grease-gun an' all an' go round an' do all the sheaves. So he did the deck bollards. Oh, he used t' look after things.

'The sides o' the boat were reinforced where the galluses were. There wuz metal platin' over the timbers. Course, I mean, you know what'd happen if the doors come up against wood! They'd smash everything in. Thass bin known t' happen on some o' the ow iron boats. Yeah, that happened on the ow *Diamond GY603*. She wuz gittin' a bit rusty an' the door come through the after side an' knocked the ow chief out o' his bunk! Some o' them very old trawlers, yuh know, you could git hold of a shovel down below when you were trimmin' coal an' you could stick that right through some o' the plates! But the ow *Pilot Jack* wuz in good nick. Yeah, she wuz a lovely little ow boat. The only thing wuz that she leaked. That coulda bin water gittin' in through where the tail-end shaft run out through the stern. I dun't really know f' certain.

'Sometimes when that come on bad weather, we used t' git under the lee o' the land an' dodge. We used t' git under the lee in Corton Roads. You could even anchor up there, so thass where a lot o' the boats used t' git to. That all depended on the skipper, what you done. An' ow Arco, he wuz just content t' jog along. Thass why we never made a lot o' money in her, but we were happy. There were one or two little drawbacks t' bein' on her, I'll admit. We used t' git bug trouble on board. We had her smoked out twice while I wuz in her. Yeah, the council people used t' come down the harbour an' do that. They used t' seal the boat up an' fumigate. They used t' bite yuh, these little ow bugs, as you lay in yuh bunk. Yeah, they used t' bite yuh an' suck. You dint really use t' feel the bite, but you used t' come up all blotchy red. Oh, they'd run all over yuh!

'There were other discomforts an' all. The roll an' toss o' the boat wuz one. when you laid in yuh bunk, you hed a bunkboard t' help yuh stay in. That wuz a long plank an' you put that along the front o' the bunk. Then you used t' stick yuh knees up aginst that an' wedge yuh backside up aginst the side o' the boat. An' thass the position you'd gotta lay in. There wun't no toilets on board; you used t' go over the side. There wun't no hot water f' washin' with neither − only what little you managed t' scrounge out o' the galley. Sometimes the ow cook'd stick a beef-kittle on for yuh. Yuh hands used t' git so sore guttin' that you used t' pee over 'em. Yeah, everyone done that. You used t' pee on yuh hands an' rub the urine in. That used t' ease 'em a bit. Another thing wuz that yuh wrists used t' git chafed by yuh oily, so you used t' wear the ow red flannel wrist-binders. Yuh neck wuz another place f' chafin', so you used t' wear a wrapper round there. That used t' be horrible when you got them saltwater boils, an' that wuz all caused by yuh oilskin chafin'.

'As soon as you'd dropped yuh fish onta the deck, you'd shoot away agin an' then git guttin' an' washin'. What we used t' do wuz put the rooker at the bottom o' the pounds because they're full o' ammonia an' that used t' keep the ice from goin' orf. We used t' put all the prime stuff an' plaice in the wings, but the rough stuff we'd hull anywhere. That wuz always the mate's job t' ice the fish. Yeah, that an' gut all the prime fish, the soles an' turbot. He used t' bleed them as well so they were fit t' sell. Course, you used t' git a little bit o' everything, dependin' on where yuh were. Cod, haddick, brill, plaice, dabs, lemon sole. Southwold-way used t' be mainly plaice. You'd trawl a lot o' them up there. And crabs. Yeah, you used t' git a lot o' them orf there. Lovely bloody crabs they were! We used t' boil 'em up as we were comin' in an' then share 'em out when we docked. You used t' git one or two lobsters an' all, but nothin' like as many as you did crabs.

'You used t' carry yuh ice down for'ad o' the fish-room. There wuz like a little hold sort o' place, where the ice used t' go in. You used t' tairke that on board when you were in harbour. You used t' go up-through-bridge t' the Ice Company an' load up. Fuellin' wuz done up-through-bridge an' all. The oil wuz in tanks, each side o' the ship. The engineer used t' see t' all that, o' course. When you hauled, he used t' stop the engine. So he did if you were gittin' the warps too near yuh arse when you were shootin'. But the skipper always hed t' ring down first. You hardly used the mizzen sail on her at all. Well, I can't remember it bein' used very much. You'd hev it up, but you never used t' work it like you did on a drifter. That just used t' be there t' keep yuh steady an' keep the ow gal's arse orf the warps when you were towin' along.

'We used t' fish mainly orf the starboard side an' I remember bein' there once in the wheelhouse, on me own, when a bloody gret cargo boat come headin' straight for me. Everyone else wuz turned in 'cause that wuz my watch. Well, I knew this boat hed gotta give way t' me, but he dint see me. He kept comin' an' comin', an' I couldn't turn the ow gal because o' the tide. She just wun't answer. I dint know what the hell t' do, so I went down below an' called everyone out. Thass how frightened I wuz! I give 'em all a shout. In the end he did see us an' he pulled up with a bit t' spare, but that wuz a bit too near f' my likin'.

'Fog wuz another thing I hated. Thass the worst thing there is. In foggy weather you see things what aren't there! If you're moving', you keep soundin' yuh foghorn all the time; an' if you're at anchor, you ring yuh bell every so orften. But that dun't matter what yuh do, you're blind. An' yit you see shapes loomin' up all the while. Course, we hed a little radio on the *Pilot Jack*, but I can't remember us usin' it a lot, except t' call the office up. On the big trawlers the blokes used t' call their wives up. They'd git their wives a trawler-band radio an' call 'em up. Yeah, they'd let 'em know what wuz happenin' an' when they were comin' in. Some of 'em used t' abuse it. Yeah, some o' these here stalky skippers did, yuh know! That'd be, "We'll be in so-and-so, ow darlin'. Git yuh knickers orf!" Oh, there wuz all that sort o' talk. One skipper I wuz with, he used t' go on the air t' the other skippers an' come out wi' all the dirty jokes he knew! Yeah, he'd hev all these jokes writ down in a notebook an' he'd broadcast 'em!

'But goin' back t' bad weather, I remember once there in the *Pilot Jack,* when we battened down an' run f' harbour. We lashed the trawl t' the side o' the boat an' we come in on a gale o' wind. Well, the ow man misjudged it. I'd never known him t' misjudge it dodgin' his way in, but he did this time. Course, thass a nasty harbour t' git into, as you know, an' we nearly hit the piers twice. The first time we hetta nip back out agin right smart. Yeah. How the hell he got round t' git out, I'll never know. But he did. The second time we come in, the wind an' tide forced us nearly onta the North Pier. In fact, I dun't think you coulda got a fag paper between us an' the North pier! Cor, that wuz a close thing!

'Like I said before, you just couldn't fish in her in bad weather. That meant we used t' hev a lot o' time in harbour. Eileen'll tell yuh how much I used t' be in — an' we still got our weekly wage. The lumpers used t' unload the fish, an' there'd be a salesman who used t' sell it for the firm. When we were in, we used t' lay in the west end o' the Trawl Dock — what people used t' call Tuttle's Corner. Thass when Eileen used t' come down an' see me. You know, that wuz afore we were married. An' thass when my green jumper went over the side! Eileen knitted me this lovely green jumper. Well, you know how funny they used t' be about the colour green. They wun't put t' sea wi' me wearin' it, so I took it orf an' that finished up floatin' past the pier heads as we went out! That wuz like my white-handled guttin' knife what I hed give me by someone. That finished up overboard an' all. An' I dint chuck it over, I can tell yuh! Oh, yeah, a white-handled knife wuz taboo on board. "We'll never hev n' bloody luck wi' that on board!" they said, so they hulled it over the side. Yeah, thass the truth, that is. I lorst me green jumper an' me white-handled knife on the ow *Pilot Jack*.

'Mind yuh, like I've said, I did enjoy bein' on her. You used t' git so many laughs. I remember ow Olly, the cook, makin' some dumplins one day. Well, his dumplins were always a bit heavy, what we used t' call sinkers. Yeah, sinkers. One day they were that bloody hard we couldn't eat 'em, so we left 'em on the plates. The ow boy got right touchy. That he did. He collected 'em all up an' he went an' chucked 'em over the side. Just as he chucked 'em over, we come fast. The warps closed up, so we knew we were on fastener. Whether that wuz fate or coincidence that happened just as he chucked these here dumplins over, I dun't know, but all the crew blamed it on the sinkers! The poor ow boy got really upset over that! Yeah, that wuz bloody weeks afore we see any more sinkers.

'On the ow *Pilot Jack* the skipper used t' fish mainly the clear ground. He used t' carry a notebook — all the skippers did — an' that wuz his bible, his record book. That told him what he'd got in a particular place at a particular time over the years, so he could always refer to it. No one else wuz allowed t' see it, not even the mate. Anyway, like I said, we usually kept t' the clear ground, so we dint have a great deal o' mendin' t' do. We worked a cowhide on the cod end t' stop the chafe an' we used t' put a dan over the side sometimes t' mark the ground if we were gittin' a decent bit o' fish. That wuz if you were after cod. See, cod like t' git near wrecks, so if you went after cod you knew you were gorn t' hev a trip mendin'. We used t' tie a basket t' the dan t' show that there wuz a wreck there an' warn the other boats. A lot o' skippers used t' tie a basket t' the dan even if there wun't a wreck there, 'cause as soon as you put a dan over that tell everyone you're on some fish an' they'll all be there after it.

'Oh, she wuz a daddy, that ow boat! I loved bein' on her. She finished up bein' sold down south as a houseboat, yuh know. She wuz a happy ship. I can't remember anything really bad ever happenin' on her. I know we hed fasteners on her, but nothin' serious. Not like some o' the steam trawlers I wuz on. I've sin the galluses pulled down t' the level o' the decks on them — you know, by the sudden strain. O' course the plates were rotten as well. Thass why it happened. But there is a terrific strain when you do come fast. Yeah, you'd feel the whole ship a-dudderin' sometimes an' you'd gotta try an' manoeuvre out of it. Occasionally one o' the warps would part. You'd got a helluva job on then. Yeah, you'd gotta heave up on the good warp an' try t' git all the gear in, so you dint lose it.

'Thass surprisin' what you trawl up from the bottom o' the sea, yuh know. Yeah, there's all sorts down there. Just after the war you often got explosives comin' up in the trawl. Mines an' stuff like that. You couldn't always see what you'd got if that wuz in with a good bag o' fish, so that always paid yuh t' be careful if you were fishin' in a place where there wuz mines an' that. We used t' git human remains come up sometimes. You know, skulls an' legs an' arms. Blast, when I wuz trimmer in one o' the trawlers, me an' one o' the other young blokes nearly killed an ow chief. We got this skull what'd come up in the trawl an' we went down the engine room an' stuck it over one o' the gas brackets. You know, they hed the ow acetylene lights in a lot o' the boats. After we'd done that, we waited down the engine-room till the ow boy come down for his watch. He hent bin oilin' round for very long, when he looked up an' see this skull a-glowin' in the dark! Cor, you shoulda sin him! He very near hed a heart attack. My mate an' me got a real rollickin' orf the skipper over that. Happy days!'

Steam and sail together in Lowestoft dock. In the foreground the steamers, and beyond them the towering topmasts of the sailing trawlers. *Ouse LT572,* 66 tons, 45 h.p., was built in 1900. George Stock skippered her on some of the North Sea voyages described in chapter seven. To port of her is *Confier LT658* built in 1910 for H. R. Boardley, and beyond her the drifter trawler *Provider LT42,* 35 tons, 25 h.p., built in 1907 at Appledore.

CHAPTER FIVE

A Boat For All Seasons

The Drifter-Trawlers

'Now up git the codfish
With his gret ow head.
He jumped on the foredeck
To git a cast o' lead
Chorus: *Singin' "Windy old weather,*
Stormy old weather.
When the wind blow,
We'll all go together." '
(Traditional — 'Haisbro Light')

The arrival of steam power in the East Anglian ports at the tail end of the 19th century led to the development of a new distinctive breed of dual purpose fishing vessel, the drifter-trawler. Drift net fishing for herring is, of course, a seasonal business, whereas trawling is not. The idea was not entirely new as there had been 'converter smacks' in sailing days, designed to go both drifting and trawling, but the herring fleets were largely geared to a pattern of laying-up in the off-season while the crews sought other berths, or jobs ashore.

With the greater capital investment in the conversion to steam, owners had a greater incentive to keep their ships earning all the year round and so from about 1908 onwards a number of steam drifter-trawlers were built in local yards.

For the most part they were steel (though there were a few wooden built), of 80-85 feet overall length, 19 feet or so in the beam, and with a hold depth of nine feet. Their net tonnage was mainly in the 40-46 tons range and they were usually powered by triple-expansion steam engines that developed anything between 30 and 40 HP. All of this meant that they were a good deal smaller and less powerful than the pure steam trawlers, and this restricted their operations to near-water fishing.

After the war the building of drifter-trawlers increased and continued right through the 1920s because both the North Sea herring and trawl fisheries were by this time in decline and the need to keep a vessel employed the year round was all the more imperative. Many of the East Anglian drifter-trawlers not only continued to go on the Padstow sole voyage from January till May, but began to stay round on the west side for the summer months as well, basing themselves at Milford Haven or Fleetwood and working Cardigan Bay, Morecambe Bay, Liverpool Bay, the Isle of Man, and the Kish and Codling banks off the Irish coast.

This fishing was, for the most part, more profitable than that carried on in the North Sea, because stocks were more plentiful and because distances to and from the grounds were not so great. With this being the case, more and more of the drifter-trawlers began to stay round on the west side for the whole year, the crews only having a few days at home in all that time. Such periods of absence, plus the high rate of unemployment among fishermen in Yarmouth and Lowestoft, convinced a substantial number of East Anglians to settle permanently round at Milford Haven and Fleetwood during the late 1920s and 30s.

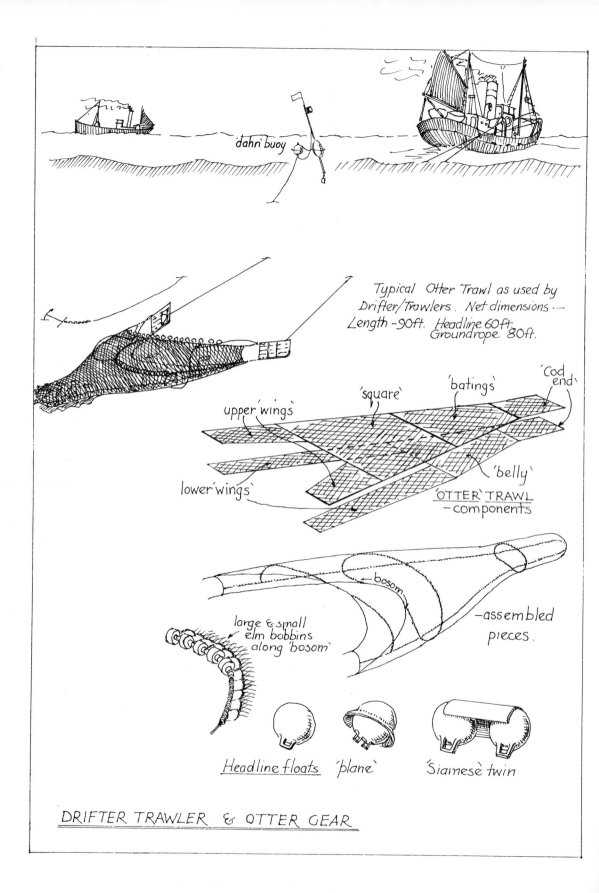

'dahn' buoy

Typical Otter Trawl as used by
Drifter/Trawlers. Net dimensions:-
Length - 90ft. Headline 60ft.
Groundrope 80ft.

upper 'wings'

'square'   'batings'   'Cod
                                end'

lower 'wings'

'belly'

'OTTER' TRAWL
- components

bosom

- assembled
pieces.

large & small
elm bobbins
along 'bosom'

Headline floats   'plane'   'Siamese' twin

DRIFTER TRAWLER & OTTER GEAR

Lowestoft was *the* port for owning drifter-trawlers and continued to place orders for them until 1931, when the last coal-burner to be built in the port was produced. This was the *Merbreeze LT253,* a 94 footer, which was built by the Richards company and later converted to diesel. She was a lovely boat, the last in line of a class of vessel that was unparallelled for the work it had to do. Whether herring-catching (with the famous Elliott steam capstan up for'ad) or trawling (with the addition of winch and galluses), the drifter-trawler was without doubt a boat for all eventualities.

Even today, 40 or 50 years after the heyday of this particular kind of vessel, one still hears the same names mentioned. Invariably, the boats talked about are the big money earners – the *Margaret Hide LT746* and *Sarah Hide LT1157,* sister-boats that were originally built as trawlers, the *Swiftwing LT675,* the *Three Kings LT517,* the *Hosanna LT167,* the *Byng LT632,* the *Patria LT178,* and the *Tritonia LT188.* The first four were part of the Jack Breach group of companies, the biggest and perhaps best-managed drifter fleet in Lowestoft between the wars, and their earnings over that period were considerable. So were the grossings of the other ones. Some idea of their money-making potential can be seen from the fact that the *Byng,* one of five boats belonging to the Vigilant Fishing Co. in the 1920s, quite often earned more at trawling than the other four boats combined did at herring-catching. A local port directory shows 84 drifter-trawlers registered at Lowestoft in 1937, but only 17 in Great Yarmouth – and most of the latter were in the Bloomfield company's Ocean fleet.

The gear usually worked by the drifter-trawlers was the otter trawl. There were two varieties of this: the earlier on-the-door gear, as it was known, and the French (or bridle) gear that began to be used in the 1920s, of which we shall hear more in a later chapter. With the former, a typical net used in the North Sea, and round on the west side, would have been something in the region of 60 feet long on the headline, 80 on the ground-rope. The headline itself was kept buoyant by spherical metal floats, while the ground-rope had its bosom weighted down with wooden bobbins, which also assisted the trawl to pass freely over the sea bed. The wooden doors which kept the mouth of the trawl open were fixed to the net by wire legs, or stoppers, four to six feet in length, and the towing warp on each side was secured to a strong steel bracket on the inside of the door. Quarter-ropes were shackled onto the ground-rope either side of the bosom, and from here they ran up on the outside of the trawl (via an eye on the headline) to the door. Once the doors were secured to the galluses during hauling, the quarter-ropes were unbent from their fixing points, led to the small barrels of the winch via sheaves on the engine-room casing and used to haul the ground-rope in.

The otter trawl relied very heavily on machinery for its operation, and much could be written about the intricacies of shooting and hauling. However, the explanations are best left to the experts, men like Herbert Doy who sailed in drifter-trawlers through the 20s and 30s, though reference to the drawings will probably help the reader to follow him:

'When you shot yuh gear, the ow man brought the boat up broadside t' wind an rung down t' stop the engines. If there wuz any wind, you always shot yuh net t' wind'ard so that'd keep the boat orf the net. You paid yuh cod end over first, then yuh belly an' batins, then yuh wings, hidline an' ground-rope. Yis, you'd git all yuh net in the water first an' all that work wuz done by hand. Sometimes we'd only work a buff in the middle o' the hidline. That wuz when we were after ground stuff, plaice an' that. If you wanted t' keep it right up, like when you were after long stuff, cod an' haddicks an' that, you'd shove some floats along it. They'd keep the ow hidline up for the swimmin' fish all right.

'Once yuh net wuz in the water, you'd lower the doors down out o' the galluses. You always let the fore one down a little further than you do the after one, 'cause you come round on that whether you're shootin' t' starboard or port. Once the doors were in the water, you'd square 'em up so they were both level. You could tell when you'd got 'em right 'cause there wuz a mark on the warps t' tell yuh. That wuz just a strand o' rope reeved through the wire. You hed one mark f' ten fathom out, one f' 50, two f' 75 and three f' 100. As soon as yuh doors were squared up, the ow man'd ring f' full ahid an' pull the wheel over so the boat run on a circular course. As you went round, you let out the length o' warp you wanted. That depended where you were how much you wanted, but you usually hed three times the warp there wuz water. So in 20 fathom o' water, you'd hev 60 odd fathom o' warp out aft.

'Once you'd got the length o' wire out what yuh wanted, a deckie would throw the messenger onta the fore warp up for'ad an' someone else would walk it inta the after fairlead an' take it t' the small drum on the winch. Soon as the drum started turnin', the messenger-hook would run along the fore-warp till that picked up the after-warp an' all an' run 'em both t' the towin' block. Once they were both in the block, you'd shove a pin through t' hold it an' reeve through the end o' the pin with a wire nettle t' stop it jumpin' out. When all that wuz done, the boat'd slow down t' about two or three knots an' you'd tow along f' whatever time you wanted.

'When that wuz time t' haul, you'd bring the boat afore the wind an' just go slow ahid. The third hand would knock the pin out o' the towin' block an' force the block open with a crowbar. Thass what we called a knockout. Soon as the warps separated, the mate'd start the winch an' heave up on the warps so they come in together. Time the doors come up inta the galluses, the boat'd be stopped an' layin' broadside t' the wind. You used t' hook the doors onta the galluses with a stopper, then you used t' unhook yuh quarter-ropes from orf the doors an' run them through sheaves on the casin' to the small drums on the winch. You couldn't use the big drums 'cause yuh warps were on there. Yeah, they were stopped, the main drums, when you were pullin' in yuh quarter-ropes. You'd unclutch them. If you were workin' bobbins on yuh ground-rope, there'd be a fair bit o' weight so you'd heave the ground-rope up as far as the rail an' then make the after quarter-rope fast t' the after fairlead. Soon as you'd done that, you heave up on the fore one till you can reach it an' then hook the gilson in it. Once that wuz in, then the bobbins could be lifted aboard.

'You'd gotta pull the net in by hand. Yeah, that you did − right till you got t' the cod end. Then you used t' shove a becket round yuh cod end an' heave it up on the gilson. You'd heave up the cod end on that an' that'd come for'ad over yuh pound. The third hand used t' stand there an' undo the knot an' the fish'd drop onta the deck. Then he'd tie the cod end up straight away an' you'd chuck it over the side. You used t' leave the wings out in the water when you hauled t' save time when you shot away agin, so once the net'd gone that only left the ground-rope. If that wuz just the ground-rope, you'd hull it over; but if you hed bobbins on, you hetta heave them over. You'd all stand there along the side o' the boat an' when she rolled, you'd shove out on the bobbins an' sing out, "Let go!" Soon as you done that, the bloke on the gilson let go an dropped the bobbins inta the water. Once they were overboard, you could unhook yuh gilson an' start all over agin.

'You were continually on the go. Yis, that you were. There'd be two of yuh on watch all the time, the mate an' a deckie, then the third hand an' a deckie. The mate an' the third hand used t' be in the wheelhouse an' the deckies used t' be down on deck, watchin' the warps. You hetta hev someone do that 'cause the warps might start t' close up an' that either meant you'd got a good bag o' fish, or you'd come fast. Soon as the bloke sung out, "She's closin' the warps!", you'd ease the ow gal down. How long you were on watch all depended on how long yuh towed. That'd be about three t' four hours on smooth ground, but praps only about two an' a half where that wuz rough. The mate an' the third hand used t' change watches each haul, but the skipper dint take a watch. The chief an' the stoker did. They used t' work from meal time t' meal time an' spell each other that way. They hed a trimmer t' shove the coal aft from the bunkers an' he used t' help on deck an' all. Anything over an hour counted as a watch, so if I'd bin down below an hour an' she happened t' come fast an' they decided t' haul, I'd hetta go on watch agin next tow. Yeah, anything over an hour counted as a watch, so you know what sleep you got if you were on a rough bit o' ground! I've sin three daylights come in round at Padstow. That I hev!

'Thass all you used t' be out round there − three days. Short trips. We used t' go round t' Padstow just arter Chris'mas an' stop there till May. We allus used t' stop so we could go t' the May Day, then we used t' go t' Milford Haven or Fleetwood. Do yuh know, I only used t' git hoom f' about a fortnight in the whole year! Yeah, arter you'd bin away 22 weeks, you'd git a week's leave. We used t' leave the boat an' come hoom by train. You got everything round at Padstow − rooker, cod, gurnets, brill. We even got seven kit o' red bream there once. But the thing you wuz really after wuz soles. Course, where they are is very rough ground, but you went prepared f' that. You could strip a set o' ground-ropes in a night round there, so you always worked bobbins. At each end o' the bosom you'd hev a bigun, so that'd keep yuh bunts up orf the ground an stop 'em pullin' out. You used t' work bullock hides on yuh cod end. Hell if they dint used t' chafe up round there. Cor, an' dint they used t' stink when they lay about in hot weather! They'd be crawlin' wi' maggots.

'That wuz all night fishin' round there. You dint do much good durin' the day. We never used t' stop an mend the trawl night time. No. If we got a split, we just used t' cobble it up an then cut it out an' do it later. Sometimes you just used t' keep cobblin' up the whole trip, lacin' the holes an' stickin' bits o' double-mesh stuff in yuh quarters. You'd hev a nice mess then when yuh got inta harbour! You'd probally hetta hev a new set o' bobbins an' all, but that dint matter. You were round there t' git soles, so you dint lose no time o' night time if yuh could help it. Another place we used t' go f' soles, when we were at Milford, wuz St. Govan's Head. Thass just orf Tenby. That wuz rough ground as well. An' there used t' be another place not far from there where you used t' git full up wi' them bloomin' crawlers. They're like a starfish, they are, an' they're right rough. They used t' play hell wi' yuh trawl.

'You want fine weather f' trawlin', but you want a little draught o' wind as well. That keep the gear orf yuh side when you're shootin' an' haulin'. If thass flat-a-calm when you're shootin', you'd hetta lower yuh doors down a little way an' steam the boat round t' keep yuh trawl clear o' the ship's bottom. If you dint do that, you might git the net caught round yuh screw. What you used t' do wuz lower the fore door inta the water an' hold it, then drop the after one down about three or four fathom. As the boat came round, the weight used t' be on that after door an' that'd pull yuh gear round clear o' the ship. You could tow any way yuh liked, but if you went an hour before the tide, that might take yuh three t' git back agin t' where yuh started. This wuz round at Milford. When you're towin' afore the tide, you can ease her right down, but when you're gorn hid at it you want twice the revs. Sometimes round there we'd lose our mark. You know, we'd drop a dan an' tow afore the tide; then when we come back up, we'd lorst the dan. See, we never hed no Decca then t' tell us where t' go.

'A lot o' the ow skippers used t' go t' the same places round the Westward every year. Ow Ted Chilvers who I wuz with, he hed it all marked up in the wheelhouse − what grounds he went to, when he went, what fish he got. That wuz all writ up on one o' the beams what went acrorss the top o' the wheelhouse. We used t' go acrorss t' Ireland an' work the Kish an' the Tuskar an' all them places − the Lucifer an' the Blackwater. Outside the Tuskar you used t' git a lot o' thornyback rooker, but when we went inta Morecambe Bay that'd be skate. We used t' go inta Morecambe Bay a lot an' work alongside the beach an' that'd be all ow blue skate there. Well, that wuz all mud orf there, see, along by Blackpool, an' skate like mud.

'If we were arter hake, we used t' run acrorss t' The Chickens. Yeah, thass the place f' hake, out orf the Isle of Man. We used t' go about sou'-west o' The Chickens an' drop down somewhere there. You used t' fix up lighter gear f' hake fishin'. Yeah, yeah. Just ground-ropes wi' nothin' on 'em. Skeleton ropes, we used t' call 'em. There wun't no dangles on, no chain or nothin'. If you wanted t' dig soles out, you'd hev chain on yuh ground-rope an' you'd work titlers as well − a bosom titler an' a door-to-door titler. But f' long stuff like hake, you dint need all that, so yuh gear wuz a lot lighter.

'There wuz some good fishin' grounds round there. Yeah. We used t' run about nine mile out from Peel harbour, Peel pier, an' drop a dan there. That wuz good fishin', that wuz, if you could git it exact. You'd git a good mixed fishin' there. The only trouble wuz you'd git in each other's way sometimes. Yis, that you would. Mind yuh, that could happen anywhere. I remember when we were orf Padstow once in the *Cyclamen LT1136* − ow Spiff Mayhew in the *Jenwil FD40,* he caught our gear. An' blow me, if the *George Baker LT1253* dint go an foul Spiff's gear! Yeah, there wuz three of us hung up, so we slacked our warps down an' let Spiff heave up 'cause he wuz a bigger boat. Well, he cleared it, but hell if there wun't a mess!

'We used t' be full up wi' gear when we went round the Westward. We'd hev about six or seven trawls an' stick 'em ashore when we got there. The different firms used t' hev these ow railway carriage things t' keep the gear in. You'd take about six or seven sets o' bobbins, an' you used t' carry a spare door each side an' a couple on the top o' yuh hatch. Then there wuz yuh spare warps − oh, you used t' take a lot o' stuff an' you used t' shove it all ashore when yuh got there. When you were fishin', you used t' carry a trawl each side an' put yuh spare gear down the fore-room. That'd be suffin like a set o' doors an' a spare ground-rope, an' all the other odds an' ends you might use on the trip.

'We used t' live on board the boats time we were round there. Yis, you'd sleep on straw mattresses! You'd take a go-ashore suit wi' yuh an' you'd keep that up in the hid o' yuh bunk. You used t' send yuh washin' hoom by post. You'd take about four or five shiftenins away wi' yuh an' then you'd send 'em hoom a couple at a time when they got dirty. If you were a good washer, you'd do 'em yuhself in a bucket. There used t' be nine of yuh on board: skipper, mate, third hand, two deckies, a trimmer, chief engineer, second engineer an' cook. That used t' be a bit cramped on board, but we dint use t' mind.

'The skipper an' the mate used t' git paid on a share. They'd git so much in the clear hundred after expenses had bin met, but the crew were all on a weekly wage. Yuh wives used t' hetta go down onta the market once a week an' draw that from the company orffice. The blokes on board never got nothin', only a bit o stockie bait. You used t' hetta trust t' that f' spendin' money. When we first went round the Westward, we got gurnets, small rooker an' rolls f' our stockie. Yeah, the rolls out o' the whitins an' cod. The salesman used t' sell all that orf an' then we got the money. Sometimes you'd do all-right; other times that'd be a bit thin.

'Mind yuh, you used t' eat well. That wuz all wholesome grub on board, but we dint used t' git no afters. No, we'd hev light duff, suet duff, beef puddin' an' that sort o' thing f' dinner — all good solid food. An' in the mornins you'd allus hev a fish breakfast. Yeah, you allus hed a fried fish breakfast. An' if you got any decent latchets, you'd hev baked latchet f' tea. You hed need eat well, though! You were on the go all the while. I've bin in on the mornin' tide round there at Padstow an' out agin on the night tide. We done that more'n once when I wuz in the *Cyclamen*. Cor, you dint half git some soles. We got six ton there once in five nights! Yeah, we got three ton an' a half in two nights, went in an' landed, an' then come out agin the nexta day. We hed two more nights out an' got another two ton an' a half. We made £700 f' them two trips. What'd they make t'day, do yuh reckon? — a multitude!

'You used t' burn about 18 t' 20 ton o' coal a week. You used t' bunker about 20 ton, an' then you'd hev about 30 hundredweight round aft, in bags. Yeah, there be 15 hundredweight each side, all bagged up. You might hev a little loose on the deck as well, but you'd soon lose it if you got a sea aboard. Yeah, I've sin coal washed all over the show! That loose stuff'd soon go out through the scuppers. When the coal strike wuz on in 1926, we were workin' out here in the North sea an' we used t' go acrorss t' Ymuiden an' git enough coal f' two trips. See, we were workin' seven or eight day trips then. We used t' git that ow brick coal. That wuz shaped like bricks an' that wuz terrible stuff t' burn. You used t' hev it right the full length o' the decks, all stacked up riddy t' use. Cor, that used t' make hard work f' the engineers! That wanted s' much draught t' burn right. Course, the engineers used t' clean the fires every watch. They used t' rake the clinker out an' the trimmer used t' pull it up the ventilator in a bucket an' hull it over the side.

'In the summertime you used t' fish in the North Sea, workin' the banks. You'd be down on the Leman, the Ower an' the Long Shoal. When you were on the Leman, you could only work yuh port side gear towin' down inside the bank. If you worked yuh starboard gear, the tide would drive yuh over the bank inta the Ower. You used t' git brill, soles, plaice, everything, down there. Thass sandbanks, yuh see, so you'd work light gear. Well, you'd hev dangles on, but you wun't work bobbins. Arter Chris'mas, if you dint go

Above: roker being de-winged on Padstow fish market. The fish are held up on a hook and the wings sliced off with a large knife. In the foreground a good haul of blonde rays lies spread out.
Below: Brixham sailing trawlers leaving harbour in the 1930's with *Ethel Lilian BM309* astern. At Brixham and Lowestoft some sailing smacks worked right up to the outbreak of the Second World War.

round t' Westward, you used t' go up the Hinder an' them places, the Gabbards an' the Gallopher. You'd praps git three or four trips in, messin' about up there, up as far as the San Ditty.

'That wuz a night-time fishin' out here in the North Sea. Yeah, you used t' git so many bloomin' gurnets durin' the day. They used t' chafe all yuh plaice t' blazes. I mean, you did fish durin' the day, but the plaice'd all be chafed t' buggery wi' these here gurnets. Night times you dint used t' git any; no, they used t' swim up then. Another thing you used t' git were these red-nosed plaice. Thass when they used t' bury themselves down in the sand when an easterly wind wuz blowin'. You used t' git that all over the North Sea, an' thass the worst wind there is f' fishin', an easterly. Sometimes we used t' go up t' the Varne. You used t' git a nice lot o' turbot an' brill there, but just orf t' the south'ard that wuz as rough as hell an' you used t' git a lot o' thornyback rooker.

'I hev worked orf the Isle o' Wight, but that ent much good round there. I've bin orf Dungeness an' all. We worked there all one week an' landed stuff on Brighton Beach. Yeah, we took the fish ashore in the little boat. Gurnets an' dorgfish, that wuz. They're hell f' dorgfish round there. These were the spotted dorgs we hed, what we called nurses, an' they made a good bit o' money there on that beach. We landed all the other stuff inta Ramsgate, all the plaice an' that. That wuz when I wuz in the *Cyclamen*. Course, I wuz in her a helluva long time.

'You dint used t' git down t' the Dorgger much in a drifter-trawler. That wuz a bit too far from Low'stoft. We went t' the Dowsin' an' worked there, though. Thass a proper rough bit o' ground; you could lose a lot o' gear down there if you dint work it right. Due south used t' be the way. We used t' work the lightship; keep that due south. That wuz all stones down there. Yeah, an' another side o' the ground wuz ross. That'd chafe a bloomin' net t' pieces, that stuff would. You used t' git a good lot o' lemon soles down there an' you used t' trawl up herrin' an' mackerel as well. They were maisy herrin'; the Dowsin' wuz known for 'em. You used t' git 'em at The Knoll an' all, maisy herrin' from the bottom. So you did round the Westward in one or two places. You used t' git a few orf The Chickens.

'You were allus away round that west side on the drifter-trawlers. Thass the grounds, yuh see; the fish wuz there. You could work all year round here too if yuh wanted. I wuz along o' Sam Read nine year in the *Cyclamen,* then he went an' sold her an' bought the *Ocean's Shield LT386*. We were all riddy t' go t' Padstow when he sold her, so we dint go away that year. We stayed hoom an' worked out o' Low'stoft. We got some nice trips down the Leman an' Ower that year. In fact, we kept down there most o' the summer. You used t' hetta clear out o' the way when the hoom fishin' started 'cause all the drifters used t' be arter herrin' then. Yit you could still git a bit o' trawlin' in 'cause the drifters dint work right on top o' the banks, where we did. Another place we used t' work wuz the Brown Ridges. The first one is about 45 mile out from Low'stoft. Then you'd git the Middle Shoal, which is about 50 odd mile. The outside one is 63 mile out, an' you wun't be far from Ymuiden then 'cause Ymuiden is only about 90 mile acrorss. That wuz a good bit o' ground out there − nice an' smooth. You could hev a trawl on f' months an' that'd last on that sand.

'When you used t' sort yuh fish out, you used t' keep yuh soles sep'rate an' yuh plaice sep'rate. Rooker you'd chuck anywhere. They'd go underneath yuh perks along o' the other rough stuff. You used t' hev a pound f' yuh rough stuff – whitins an' dabs an' that sort o' thing. We used t' git ling an' conger round the Westward an' they'd git chucked in along o' the rooker. The soles were the things. You used t' hetta squeeze the blood out o' them an' pair 'em orf. You'd wash 'em round in the sole tub an' then pair 'em orf t'gether, back t' back. Turbot an' brill you used t' nip the blood out of as well, but all the other stuff you'd just wash round in baskets. You know, you'd git a sole tub full o' water an' just swill the fish round in a basket.

'You used t' hetta clean yuh holds out every trip. Oh, good God, yis, you hetta do that! That wuz git yuh fish out, scrub out everywhere, ship the pound boards all up, go git the ice an' go git the coal. Time you'd done all that, that wuz dinner time. Away you'd go agin next mornin' at eight o' clock! We used t' hetta basket the ice up sometimes an' hump it aboard. The ice wherries used t' come alongside yuh an' you'd cart it down inta the fish-hold, t' the ice-locker. There wuz 32 baskets t' the ton an' we'd sometimes ship six or seven tons on board. The ice-locker wuz a wood one an' that wuz draught-proof 'cause a draught'l cut yuh ice away like blazes. You used t' hev about three or four cookins o' meat in there along o' the ice, so that'd keep nice an' fresh.

'You were allus on the go a-trawlin'. Yes, that you were. Night an' day. Round at Padstow I've laid down on the floor o' the cabin wi' me oily on! That wun't worth turnin' in. Time you git yuh gear orf, that wuz time t' put it on agin. Thass the truth, that is. You can ask any ow fisherman that about Padstow an' he'll tell yuh. Ow Bill Dawes made up a song about it – "The Sleepy Valley". Yeah, that wuz just orf Newlyn Rock, the Sleepy Valley. You never got no sleep there 'cause you were on the go all the while. That wuz a rum job, ow boy, I can tell yuh that. That wuz work. I've done 72 hours in one stretch round there. Thass like when the *Ocean Sunlight YH28* come in there once. The skipper went an sent f' the doctor. He say, "My crew ha' got the flu." After the doctor had looked at 'em, he say, "There ent nothin' wrong wi' your crew. They're just tired out. They want a rest." See, they were messed up. They'd bin workin' f' so long.

'The skipper used t' be in the wheelhouse practically all the time, when he wuz up an' about. Mind yuh, that wun't exac'ly easy for him 'cause he'd gotta see arter the ship time you wuz shootin' an' haulin' an' that. That wuz his responsibility. The mate's job wuz t' see that the fish wuz iced away prop'ly an' the third hand's job wuz a-washin' on 'em in baskets. The deckies used t' hand 'em down below, an' the trimmer used t' be down there along o' the mate t' break the ice up. That wuz crushed ice, but sometimes you'd need a pick t' break it up with. Moostly, a shovel'd be enough t' do it, but just occasionally you'd need the pick.

'When you landed, that dint make much difference how you got 'em out. The first stuff out wuz the freshest, but that dint make much difference t' the price 'cause you were only out about three or four days. Some o' the boats used t' go on the smooth ground 14 or 15 miles out orf Padstow, an' they'd be away about a week, but we used t' work the rough ground about two or three mile from land. You used t' be nearly alongside the land there at what they call Man an' His Man. Thass just betwin Trevose Hid an' the Godrevy Light. There's two rocks there, an' they used t' call it Man an' His Man. You used t' tow till you closed 'em up, then you'd pull around an' tow back agin till you opened 'em up.

'You used landmarks a lot when you were trawlin'. There used t' be a nice bit o' smooth ground round there, an' that wuz the one orf Spion Kop. That wuz a big hill an' you'd sight that an' drop a dan. Very orften you'd hefta pick the dan up an' steam five minutes t' clear the rough. Then you'd drop it down agin. That wuz all lead-line in them days t' fine yuh depth o' water; you dint bother about findin' what sort o' bottom it wuz. You knew that alriddy − slate! Years ago, out o' Low'stoft here, they used t' use the ow 14 pound lead t' test the ground. They'd put a little bit o' grease on the bottom an' pick up the soil. When they worked the Gut o' The Knoll, they used t' chuck it over an' see if they could pick up any pipey. Pipey weed. Thass what plaice live on. Yeah, the gut of a plaice'd be full up wi' that.

'The bobbins you used t' work round at Padstow were three sizes: 12 inch, 14 inch an' 18 inch. The rougher the ground, the bigger the bobbins you used. Out here in the North Sea you moostly worked the 12 inch ones 'cause the grounds wun't so rough, but round there you needed the other ones as well. You dint work titlers at Padstow 'cause they'd allus be breakin' on the rocks. That wuz hard on the doors an' all. You'd allus be bustin' them. Sometimes you'd clout the rocks so hard that when you hauled all that'd be left wuz the angle irons! Mind yuh, the otter gear wuz a good way o' fishin'. You hed the doors t' keep yuh trawl open, an' then you hed them legs t' join the net t' the doors. Sometimes they were about four foot long, sometimes six − that all depended. But they were a good idea 'cause they let all the ow muck on the bottom pass through without tearin' the net. When you were towin' along, all the shit used t' go through the legs, so you wun't split yuh net.

'We got caught poachin' once. I wuz along o' Sam Read then. I wuz in the *Robert Gibson LO441,* a trawler out o' Milford. We got picked up by the Irish bogeyman an' took inta Ballycotton. They come aboard an' took the gear an' the fish what we'd caught. We couldn't do nothin' about it; we'd got caught fishin' inside the limit. They even took the warps orf the winch. Well, they left us one 'cause the windlass wuz all rusted up an' we had t' have somethin' t' drop the anchor with. You could do that with a trawl warp, so they left us one on the winch. Soon as we'd got out, we pulled that warp orf the winch, halved it an' put half on each drum. Then we rigged up a trawl from our spare gear an' went an' shot in the same place again! We hed plenty o' spare gear left; they only took what lay along the side o' the boat. They dint touch what yuh hed stowed away below. We got a nice trip there − cod an' plaice an' all sorts.'

Steam Drifter/trawler. c.1930
45 tons net.

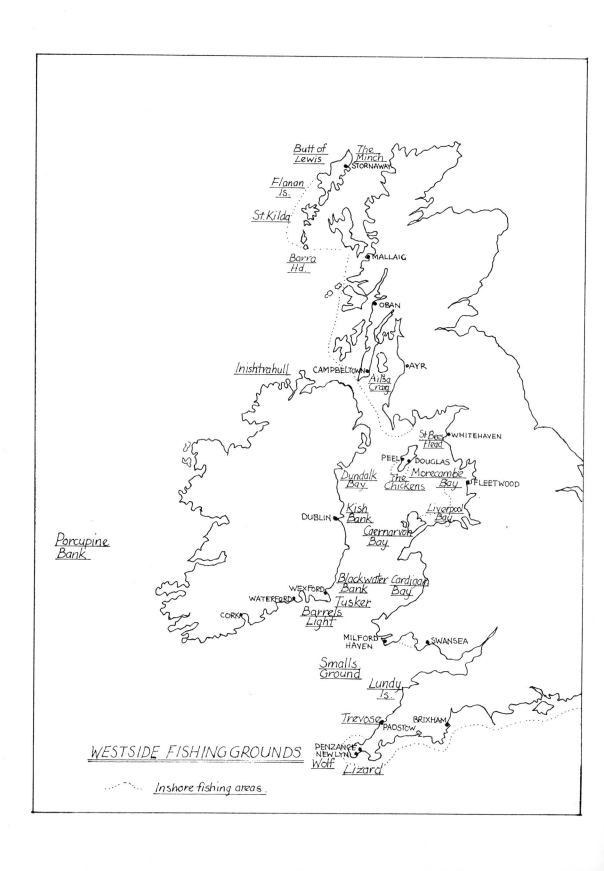

WESTSIDE FISHING GROUNDS

.... Inshore fishing areas.

CHAPTER SIX

West About

Fun and Fishing Away From home

'Throw out the life-line across the dark wave,
There is a brother whom someone should save;
Somebody's brother! Oh, who then will dare
To throw out the life-line, his peril to share?'
(No. 772, Sacred Songs & Solos – Ira D. Sankey)

The exiled life of the drifter-trawlers based in West Country ports was the common experience of many East Anglian fishermen of Herbert Doy's generation. It was a migratory way of life that came to an end only in the 1950s, as the last of the steam boats gave way to their diesel successors. The increased range of the new generation of motor trawlers and drifter-trawlers enabled them to concentrate on the North Sea, going down to grounds off the Danish and Norwegian coasts, as well as the banks out from Aberdeen, while still based on their home ports. Those areas which are thought of today as being the Lowestoft boats' traditional grounds (the Monkey Bank, the English Klondyke and all the rest) are, in fact, only 25 to 30 years old in their large-scale exploitation by the Suffolk port.

Yet with North Sea hauls now on the decline again, there are moves afoot to see whether some of the grounds off Cornwall and in the Irish Sea could once more support a viable fishery by Lowestoft vessels. Thus do wheels come full circle! With modern transport the crews will doubtless do better than two annual weeks at home, yet life for the old timers on the voyage west about was not without its lighter moments as Ned Mullender recalls in some of his reminscences of Padstow and its famous May Day ceremonies.

'I wuz along o' my father afore the First World War an' he took the *Glen Heather LT62* new in 1913. We went herrin' catchin' up till Chris'mas, then in January she went down t' Grimsby t' be fitted out wi' the galluses an' winch f' trawlin'. When we fetched her back agin, we done two or three trips out here, out o' Low'stoft, an' then we went round t' Padstow arter the soles. You hoped t' make a better livin' out o' that; thass why you went away. We went away in the February o' 1914 an' we were round there till May. Then we come hoom an' got riddy f' the Shetland herrin' voyage. We just took the fore-galluses orf f' that. We left the winch an' the after-galluses on 'cause they wun't in the way.

'When we got back, we did a bit o' trawlin' afore we went t' Scotland an' my father an' me both come over queer. We hed different doctors an' when we went an' see 'em, they said my father'd got influenza an' I'd got a sore throat. I alriddy knew that! Anyway, my father stayed at hoom, but I went back a-fishin' 'cause I dint feel too bad. When I come in agin I did! There wuz three or four of us runnin' t' catch a tram up t' Pakefield, an' my feet fared as though they were a-goin' like hell an' a-gatherin' no ground. In the end the others hetta pull me on board the tram. "Whass wrong wi' you?" they say. "Hell if I know," I say. Anyway, t' cut a long story short, that turned out both my father an' me hed got typhoid fever! Dun't ask me how. I couldn't tell yuh. The only thing I can think of is that just afore we come hoom from Padstow we hed an ow liver-jar explode. That wuz a wooden barrel thing f' storin' cod livers in an' that wuz standin' on the quay when the bung blew out. All the livers inside hed fermented an' there wuz this stuff a-foamin' out all over the quay. Talk about stink! My father an' me were the ones what washed it orf the quay, an' I orften wonder if that ent what give us the typhoid.

'Course, that Padstow voyage wuz a very big thing f' the Low'stoft boats. My father hed bin goin' round there f' donkey's years. Even some o' yuh top herrin' men went round after the soles. Oh yeah. Ow Cronjie Capps took the *Bluebell LT1044* round there about 1910. Now, she wuz one o' Tom Thirtle's drifters, a wooden boat, but Cronjie took her round with a beam trawl on each side. I dun't know how he got on, but you could work a beam trawl orf a steam drifter all right. After the war a lot o' the drifter skippers went in f' trawlin'. See, they were herrin' skippers really, but they were in drifter-trawlers an' when the trawlin' season come along, well, they still kept skipper o' the boat. Course, if you go back far enough, drifter skippers dint even hefta have a ticket. Trawler skippers did, though, 'cause there wuz a lot more seamanship needed in trawlin'. Well, you can imagine what happened when you got blokes who dint really know what they were doin'. There wuz some panickin' stations then at times! One night some feller come an' shot his trawl right over the top o' our fore door. We were haulin' an' his fore warp come over the top o' our blinkin' door. He just chopped acrorss us.

'The length o' season round there at Padstow varied. That all depended on when yuh went an' when yuh left. Sometimes you'd leave Low'stoft in January, sometimes February. You'd usually stop round there till May, but I remember bein' there one year inta June. Then I went onta Fleetwood, an' stayed there till the hoom fishin'. I liked Fleetwood. That wuz all right up there. Yeah, plenty o' skipper's stockie bait! Specially the ow scollops. Tanner organs, thass what they used t' call 'em round there. We dint hev a bad little fishin' that year. That wuz about my second year as skipper, I think − 1928, or suffin like that. I know we got a telegram from the owners, which said, "Plenty o' soles at Lundy", so we went back down t' Lundy an' worked there for a bit afore we come hoom.

'Durin' the 1920s you hed about 60 or 70 ships round at Padstow out o' Yarmouth an' Low'stoft. Yeah, some o' the big names used t' git round there − Arthur Artis, Oscar Pipes an' them. A lot o' the wives used t' go down t' Padstow an' all. Not where there wuz a large fam'ly o' course, an' that'd be just the skippers an' mates wives. Well, the rest o' the crew couldn't afford it. I hed two children, a boy an' a girl, an' we used t' take rooms in a boardin' house sort o' thing. The wife an' me hed a bedroom, the kids hed a bedroom, there wuz a lounge an' we hed the use of a kitchen. My boy even went t' school down there, but my gal dint 'cause she wun't old enough. One thing what helped the skippers t' keep a fam'ly round there wuz the discounts they used t' git on groceries an' coal. Whoever you dealt orf f' them used t' give yuh a discount f' yuh custom. An', o' course, the ow chief engineer used t' git a little backhander orf the coal merchant too.

'When there wuz gales o' wind, the harbour used t' be full up wi' boats. Like I said, there musta bin 60 odd round there alt'gether. See, Bloomfield's fleet from Yarmouth wuz there as well as all the Low'stoft boats. There wuz wooden ones along o' the steel ones, yuh know. Not all that many, but praps seven or eight. One or two o' Jack Breach's went. The *Eileen Emma LT342* did; I can remember her bein' round there. In the 1930s a lot o' the ships started t' use Fleetwood more'n they did Padstow 'cause the harbour expenses wun't s' much round there. Where you might spend 70 odd quid a week at Padstow f' harbour dues an' landin' dues an' all the rest of it, that'd only be about 50 at Fleetwood. Well, that meant you'd save 20 quid on yuh expenses. An' another thing, the cost o' labour f' repair work wun't s' dear up at Fleetwood neither.

'When I wuz skipper o' the *Impregnable LT1118* f' ow Cutty Robbens, I think my money wuz about three quid a week. I ent absolutely certain 'cause, I mean, I'm goin' back now t' 1928, but that wuz round about three quid. That wuz an allotment 'cause you were paid on a share, an' they'd deduct that from yuh earnins at the end o' the voyage. The mate wuz on a share as well an' he got about 50 bob allotment. The chief engineer got about 50 bob as well, but that wuz a weekly wage wi' him. The rest o' the crew got a wage as well. I think the third hand's wuz about 35 bob or a couple o' quid an' the stoker's about 30 bob. Then you hed yuh two deckies; they'd be on about 25 or 30 bob. The trimmer, he got 17/6d odd or a quid — I f'git now — an' the cook, he'd draw about 15 shillins or a bit more. If he wuz a full-grown man, not a boy, he'd probably git about the same as the deckies. Everyone got stockie bait, lobsters, gurnards, crayfish an' that, an' the skipper got all the small rooker as his perks. The skipper's share wuz about £10 on the clear hundred, after you'd met all yuh expenses, an' the mate's about £7-10-0d. The skipper used t' draw the stockie bait an' share it out, an' o' course some of 'em used t' try an' fiddle the blokes. Ow Cutty Robbens wuz a good man t' work for 'cause he used t' pay all his men a poundage on top o' their wages. Ten bob in the clear hundred is what you used t' git.

'Course, in them days you used t' look after yuh fish. Not like they do t'day — just hull 'em anywhere! For a start, you'd sort 'em all out inta different pounds. Dabs, whitins an' lemon soles used t' be put t'gether, plaice an' haddicks you kept separate. Then yuh prime fish, yuh soles, turbot an' brill, they used t' be kept separate. Well, you nursed them 'cause they were worth a bit o' money. You hed t' be very careful wi' the turbots 'cause they're very easy t' mark, an' once they are marked they lose a lot o' their value. You used t' bleed them by stickin' yuh knife in where the backbone run inta the tail. You'd pierce there an' all the blood would run out. Yuh cod you used t' shelf — if you hed the shelfin' room. You'd hev two or three pounds an' you'd keep puttin' the ow shelf boards one on top o' the other till you built right up. They were thinner than yuh pound boards an' you used t' lay yuh cod belly down on them on top o' a layer o' ice. There wuz just a single layer o' cod each time an' they used t' keep beautiful like that. The rooker used t' go in the bottom o' the pounds t' make a bottom layer f' the other stuff, an' then when you'd done that you'd put 'em in yuh wells. They were the pounds right underneath yuh perk boards.

'Sometimes in the ow drifter-trawlers, if you got a lot o' wind an' tide, you hetta give up towin' along. The boats hent got enough power t' keep goin' in a lot o' swell an' that meant yuh doors would drop. Soon as that happened, yuh trawl wun't fishin' right so you'd haul it in. What way you towed all depended on where yuh were. I mean, you couldn't tow afore the tide all the while; you'd be miles away orf the fish if yuh did. There were times when you hetta go back up agin an' that wuz hard work f' the ow gal. What you liked t' do, if you could manage it, wuz t' work one tide goin' one way an' then come back up agin on the next tide. But you could only do that if there wuz a bit o' fish all the way down that line, see. When you hit the soles round there at Padstow you knew it! I recollect one time that we got 11 quarter-cran peds o' soles in a three hour drag. Heaped up peds at that, about nine stone o' fish in each one. Yeah, we got about three ton alt'gther that trip in about 48 hours. Sometimes there'd be so many on the market round there that they wun't make n' money! I mean, I've sold soles at 3½d a pound. Yeah, that I hev. An' f' every shillin's worth o' fish you sold the buyer used t' git a farthin' discount knocked orf, so that meant you actually got 11¾d.

'There wuz one little bit o' ground round there I used t' like workin'. That wuz about 14 mile out from Trevose Head an' you'd git some good fishin' there. Course, you'd git three times the amount o' fish in the dark than what yuh did durin' the day, so what yuh had t' do at night wuz git yuh bearins as accurate as yuh could. You couldn't see yuh landmarks like yuh could durin' the day, so you'd work on the lights. You'd git the bearin' o' Trevose an' the bearin' o' Godrevy an' the flicker o' Pendeen. You couldn't actually see Pendeen Light, but you could see the loom of it comin' round, an' you'd git these three bearins, see, like a three-point bearin' an' drop a dan down. You'd hev a light on the dan so you could see it in the dark an' you'd work t' that. Course, you only towed about two hours, but that wuz enough t' git several weighins o' soles. And in them days we used t' cooperate with each other, us skippers. Well, not all of us, but a lot did. Oh yeah, we used t' give one another the news of how we fished an' that sort o' thing. We'd go an' hev a drink at dinner time in the pub, see, an' hev a yarn. There wun't s' much keepin' quiet then, not like there wuz after the war. No, you'd more or less give one another information at one time o' day.

'Course, we all used t' hev a good day t'gether on May Day. Yeah, that we did. I hed a rare ow day there one year! A lot o' the boats hed alriddy left, but there wuz still several of us about. This'd be about 1928. I know I started drinkin' about seven o' clock in the mornin'. One o' the buyers round there wuz a Low'stoft man who'd made his home in Padstow an' he invited me an' my young guvnor, one o' Cutty Robbens's boys, inta his orffice f' a drink. Well, bor, we'd got through a bottle o' port an' a bottle o' whisky afore sellin' time. That wuz eight o' clock, incidentally. An' o' course we sold the fish then an' started t' git prepared f' the hobby-horse business.

'I went back t' the boat, the *Impregnable*, 'cause this wuz a year when the wife an' the children dint come down. My cook, Billy Nicholls, wuz stood there. He come from Tunstall, down near Leiston; his father wuz shepherd on a farm. I say to him, "Billy, everybody's goin' t' hev a day orf t'day. You hent gotta do no cookin'." Well, after we'd got everything cleaned up an' squared up riddy t' go t' sea agin, the pubs were open, so we adjourned inta the pubs. See? While we were settin' there in one, along come a police inspector who I used t' save a bit o' fish for. I went back t' the boat with him t' git him his fish an' when we got there, the only one o' the crew about wuz the chief. He'd blew down an' wuz cleanin' the back ends o' the boiler out. After he'd got the inspector his fish, he come inta the cabin, where I wuz, an' he say t' me, "I think you've hed a good day. You'd better lay in."

'Well, I dint feel like layin' in, an' on the cabin table wuz a plate o' cooked ham an' chicken. "Where did that come from?" I say t' the ow chief. He say, "The cook o' the Metropol Hotel sent it down." "Good," I say. "Git out the mustard pot an' we'll hev some.'' So we did. After we'd finished, I decided t' go an' hev a wash an' shave. Well, that got him worried 'cause, I mean, I wuz well away an' I used a cut-throat razor! Anyway, he stood there an' watched me time I shaved, t' make sure I dint cut m'self, I spose. Soon as I'd done, I went t' the cupboard t' git me best rig-out so's I'd look smart. We hed this cupboard what wuz meant t' be a pantry really, but we had all our portmanteaus in there with our clo's in. I got dressed up right smart an' wuz all riddy t' go. The chief wun't come, though, 'cause he hent finished what he wanted t' do, so orf I went on me own.

'The first thing what happened when I got ashore wuz that I runned inta a barrer full o' chocolates an' sweets an' oranges. There wun't no one with it, so I got inta the shafts an' drew it round the corner. There wuz a whole lot o' kids a-follerin' me, so I went an' give everything away − all the sweets an' that. Course, I got caught in the act by the man who owned the barrer! I say to him, ''Whass the damage? The kids ha' enjoyed it.'' An' that wuz that. I settled up fair an' square. A little while later I met up wi' some o' the blokes an' we goes orf up t' the fair. When we gits up t' the fair, there's nobody about. Thass May Day, see. Everybody's down in town. Mind yuh, we did hev some kids follerin' us an' there wuz a bloke there, standin' about near one o' these roundabout things. ''Will yuh give the kids a run?'' I say to him. ''No,'' he say. ''Go on,'' I say. ''I'll pay yuh, as long as yuh dun't rush me too much.'' He say, ''All right. I'll run it f' ten shillins.'' So he did. The kids thought that wuz marvellous, an' after they'd hed two or three rides on this here roundabout away we come.

'When we got back inta town agin, the others buzzed orf inta the pub an' left me, so I wuz all on me charley own. I wuz still feelin' pretty merry, so I thought I'd go an hev a look at the May Day procession. Well, bor, I dun't know how it happened exac'ly, but I got at the front of it along o' the hobby-horse. Yeah, I wuz up there a-playin' a kittle-drum. I'd picked this kittle-drum up from somewhere an' there wuz the ow hobby, a-dancin' out in front with a bloke playin' the pianner accordion. Oh, I enjoyed myself! I even played kittle-drum with the two sticks on a policeman's hat. O' course, I apologised afterwards; I knew I'd made a mistake. But the policeman took it all in good part, an' then I even hed m' photograph took wi' the sergeant an' the inspector. I've still got the pictures, only they're so faded now you can't hardly see 'em. Thass one May Day I'll never f'git.'

Those snapshots may have faded a little with time, but Ned's memory remains crystal clear. Equally graphic are the recollections of Horace Thrower (born 1904). He came to drifter-trawling rather later and his experiences span the period during and after World War Two:

'I first went trawlin' the fore part o' the war, when I went round t' Milford Haven an' joined the *Constant Star LT1158*. I'd always bin herrin' catchin' afore that. Well, as soon as ever I got round there, they called me up! I hetta go t' Swansea f' a medical. Well, when I told 'em what I wuz, they said they dint want me an' told me t' go back t' Milford. ''Carry you on,'' they say. ''We'll call yuh when we want yuh.'' So back I go. Do yuh know, I got two more calls like that afore I finally got called up! Yeah, you talk about the way people mess yuh about.

'We used t' fish all over from Milford. Yeah we used t' go round t' Padstow an' fish orf there. We used t' do Cardigan Bay, Liverpool Bay an' all them places. Run acrorss t' Ireland as well. Sometimes, durin' the day, we used t' fish f' cod in what they called Douglas Hole. Thass betwin Milford Haven an' the Isle o' Man. Yeah, an' soon as ever that come in dark, we used t' haul the trawl, put all the heavy gear on an' go after soles. Cor blimey, dint we used t' git some soles there an' all! One o' the boats what wuz with us, she landed over 150 kit o' soles one trip. Thass what yuh call catchin' soles, ent it? An' when we used t' go on the market an' sell 'em, they'd fetch about thrippence a pound. Thrippence t' sixpence they'd be. Well, you were lucky if yuh got sixpence for 'em. There's a difference now, ent there? An' another thing at that time, the fore part o' the war, wuz this: if you landed any blinkin' ow whitins, the gover'ment used t' take them afore anything else. Yeah, all the ten stone kits o' whitins would be gone orf the market the very first thing in the mornin'. They used t' go t' the hospitals f' the wounded, I believe, 'cause they're easy t' clean an' cook. Well, you steam 'em, dun't yuh? Anyway, the gover'ment took 'em. They wun't touch the soles.

'Now soles are a night-time fish. Durin' the day they bury theirselves in the sand more or less. So what yuh do, whenever yuh go for 'em, is put plenty o' chain on yuh ground-rope. Yeah, you'd wrap that all round so that dig inta the soil. If that wuz rough ground, you used t' hev bobbins on so you could git over the ground better. F' cod you used a lighter gear 'cause they're a swimmin' fish. Sometimes, in Douglas Hole there, we'd git so many cod that they'd start t' git out o' yuh trawl! Well, yuh know what I mean — you'd hev a real bagful. You used t' hetta make three or four bags of 'em t' git 'em aboard. Yeah, you'd git the ow beckett round the cod end, hoist up, drop that lot onta the deck, an' then do the same agin two or three times till you'd emptied the whole lot out.

'If you were an engineer, like I wuz (see, I wuz stoker), you could help t' haul when you were trawlin' if yuh liked. You wun't forced to. But if they got a lot o' fish, praps you'd pop on deck an' give a hand. You'd give a good oil round first, you know, an' fire up, then you'd go on deck f' half an hour. I wuz lucky wi' the boats I went in; they were all pretty easy-steamin' boats. The *Tritonia LT188,* what I wuz in after the war, she wuz a lovely steamboat, she wuz. You could burn anything in her. We used t' go t' Milford an' git that right small coal what they hev there. You know, what they call nuts. Yeah, that wuz the coal t' burn. You'd fire very lightly wi' that, just throw a little on at a time an' spread it over the furnace so it lay light. You'd git very little clinker when yuh did that. Praps you'd go a couple o' watches afore yuh hed t' clean yuh fires. She wuz a two furnace job, the *Tritonia,* so you could clean 'em one at a time. Yeah, that wuz quite easy t' do an' you dint lose right a lot o' steam neither.

'When I wuz in the *Constant Star* durin' the war, we used t' fish a lot orf the southern part o' Ireland. We used t' go in close there orf Rosslare after blond rooker. The water wuz so shallow that yuh blinkin' trawl would be nearly on top o' the water. We used t' go in so close sometimes that you could see people walkin' about on the shore. Yeah, you'd go in as close as you dare go, but you used t' hetta keep a good lookout f' the fisheries patrol boat. Yes, that you did. They were buggers round there orf Ireland. They'd soon hev yuh. What we used t' do wuz put a sack over the top o' our numbers in case they were about. The blokes on the lightship there were very good to us, though. That wuz the

Blackwater light vessel. What they used t' do wuz hoist their mains'l if they knew the bogeyman wuz about an' that used t' give us plenty o' warnin'. Yeah, soon as ever we saw that mains'l up, we knew exac'ly what that meant, so we'd git back outside the limit. You talk about the tricks what used t' go on!

'I know we were fishin' round there once orf Rosslare an' we got a blinkin' live torpedo in the trawl! An aerial torpedo, that wuz. We were haulin' the trawl an' all bubbles started comin' up under the ship. The skipper say, "What the devil ha' we got here?" Course, as soon as we see the trawl come up, there wuz the blinkin' tail of a torpedo stickin' out o' the net! An' that wuz a live one. Leastways, we reckoned it wuz! Yeah. So we cut everything away. We let the lot go. Yeah. Well, we darsn't do anything else. That wuz most likely alive, see, an' that couldn't ha' bin there long 'cause we'd heard the planes the night before comin' over the top of us. Well, thass where we reckoned it come from.

'Another thing round on that west side wuz the weather. Cor, there used t' be some rum ow weather round there! Specially at Padstow an' Milford. Milford wuz the worst. I remember two times durin' the war when the weather wuz so bad that we hetta come inta Milford Haven stern first. You did that so yuh bows would break the sea. Oh, thass a wicked place round there b' The Smalls. The blinkin' sea used t' run up above yuh like blinkin' mountains! You'd just hev the engine tickin' over an' you'd ease her in on the tide. Yeah, you used t' git some rum seas round there, specially with a sou'-west wind. Thass the worst wind round at Milford; that make the sea run in there somethin' terrible. An' yuh git s' much swell with it. That ent like the short seas you git out here in the North Sea; round there thass just one big roll. You could hev a boat alongside yuh, praps only 30 odd yards away, an' you wun't see him when he went down in a trough.

'If you were a good way out from port, you'd dodge it if that come on bad. Or praps if you could run in anywhere an' git under the lee o' the land, you'd do that. Course, the bigger the boat, the longer you could keep a-fishin' in bad weather, but in a drifter-trawler once you got up t' about force eight that begun t' git a bit dangerous. Yeah, thass how I think the *Lord Haldane LT1141* wuz lorst. I think she'd got her trawl down in bad weather. She wuz lorst with all hands; they never see a sign of her. I think she wuz fishin' round Swansea way when she went down. I know when we got inta harbour we heard that she wuz gone. An' we put two an' two t'gether an' we said, "There you are. Thass what come o' fishin' in bad weather." Course, we dint know f' sure, but I should say thass what happened. There wuz a lot o' wind that night − well, that lasted 24 hours alt'gether − an' if you are a-fishin' in that sort o' weather, you're practically at a standstill. Yuh blinkin' engine is goin' blinkin' full an' yit you're practically stopped. See, yuh trawl is like an anchor when thass down an' you're more or less blinkin' helpless if anything happen.

'I fished f' about six or seven months durin' the war, then I got called up. The *Constant Star* stopped round in Milford, I think. Yeah, I'm sure she did. I dun't think she came back t' Low'stoft. I'm sure a bloke what lived at Milford Haven bought her afore the end o' the war an' kept her round there. After the war I went drifter-trawlin' along o' Doff Muttitt in the *Tritonia*. We used t' work out o' Fleetwood a lot then. We used t' be out about seven or eight days, I spose. Yeah, that'd be about it 'cause yuh coal wun't last any longer. We used t' bunker about 30 ton an' then we'd carry a deck cargo sometimes, if that wuz fine weather. Yeah, we'd hev about three or four ton each side, loose. Or that might even be in bags. That wuz that there fine Milford coal what I wuz tellin' yuh about. Cor, that wuz lovely stuff t' burn.

'We used t' go after soles an' cod just the same as we did durin' the war. Yeah, we'd go in Douglas Hole just the same. An' we used t' run down t' Padstow an' all. That'd be f' soles. There used t' be good fishin', an' there used t' be a good lot o' boats too. Cor, when I think o' all the boats what used t' be there in the early part o' the year! An' all after soles. When yuh went past one another, you'd think you were goin' t' git caught up in each other's trawl. Sometimes yuh did, but usually you just about went clear. Oh, there used t' be a lot o' Low'stoft an' Yarmouth men round there just after the war. There'd be all the "Ocean" boats there – you know, the *Ocean Sunlight YH28* an' them. That wuz one o' their main voyages, that fleet.

'When you were workin' out o' Fleetwood, sometimes you'd come in with a good trip an' you'd praps git a phone message from the owners – "Go t' Milford." That meant you could git a better price there, so orf you'd go. That wun't too bad a steam t' make an' o' course you'd always be workin' somewhere betwin the two ports anyway. Yeah, you'd be in Douglas Hole after cod an' soles one time. Then, another time, you'd be down orf Milford there in the deep water, after hake. You'd draw 50 or 60 fathom there an' yuh warps would be quite steep orf the end o' the boat. Yeah, you could hardly git yuh finger betwin 'em where they come out o' the block; there wuz that much water above the trawl.

'A lot o' the boats round on that west side fishin' used t' git t' the Padstow May Day if they could. Oh yeah, everyone used t' make a practice o' goin' t' May Day. They used t' hev a rare ow day there that day. Oh, that wuz a blinkin' lovely day, that wuz. Cor, yes! They'd all dance in an' out o' each other's houses, yuh know. Yeah, that they would. Everyone'd be celebratin' all day. O' course, Helston Floral, that wuz more or less the same sort o' thing. Yeah, I went t' that once or twice when I wuz round the West'ard. Oh, everyone hed a rare ow go there. Funny how yuh got them two things in Cornwall, ent it?

'When we were workin' Fleetwood after the war, we used t' work Cardigan Bay, Liverpool Bay, Morecambe Bay an' all them places. You used t' git a lot o' thornyback rooker in Liverpool Bay at certain times o' the year. The only trouble wuz you used t' git s' much trash in yuh trawl as well. You're never sin such stuff as we used t' git there – blinkin' bedsteads an' God know what! Yeah, everything you could mention, an' all household stuff. Chairs an' God know what. I spose that'd all got washed out from the Mersey. Yeah, an' you used t' git the rooker in amongst all that. You knew what t' expect when yuh went in there. That dint tear yuh net all that much, not even a bedstead, but that wuz still a nuisance. The things used t' be a blinkin' job t' git out o' the net. They'd git all wound up in the meshes.

'I'll tell yuh another thing we once got in Liverpool Bay too. I spose that come orf a liner. That wuz all rolled up in a bit o' paper an' someone musta threw it overboard. That wuz one o' these here dirty books. You never see such language in all your blinkin' life! That couldn't ha' bin in the water long, I shouldn't think, 'cause you could read it all right. One o' the blokes took it down the engine room an' dried it out 'cause we wun't sure what it wuz at first. Once that wuz dry we soon found out! I tell yuh, you're never sin anything like it in yuh life. You wun't think a book like it would be printed. That wuz a lot

o' blinkin' wickedness, there you are! That wun't printed in this country, I'm damn sure
– not language like that. That come from America, I reckon, an' someone hed hulled it
orf a liner afore that got inta Liverpool. Well, thass what I reckon anyway.

'You used t' git some lovely plaice round there on the west side. Yeah, that yuh did.
When you were orf betwin Cardigan Bay an' Ireland, there wuz some very nice plaice
there. Another thing you used t' git when you were orf the Irish coast wuz conger. An'
they were blinkin' conger too! Cor, they were a blinkin' size. They can be very nasty, yuh
know. They bite. Yeah, we used t' keep them in a pound, sep'rate. I've sin the time when
we'd bin out praps three or four days an' the buggers were still alive! Well, they were
wrigglin' about in the blinkin' pound, anyway. They were the biguns; the small ones died
fairly quick. Them biguns, yuh know, they'd be a good bit thicker'n yuh arm. Oh, they
were vicious ow things, they were. You used t' git them on rough ground. So yuh did ling.
You used t' try an' keep t' the edge o' the ground if yuh could, but sometimes the tide used
t' carry yuh further acrorss. Then praps you'd come fast. When you did that, you'd heave
up first on one warp an' then on the other t' ease the trawl orf what wuz holdin' it.
Sometimes you'd gotta praps heave up on one warp alone. Whatever yuh did, yuh net
would orften be in ribbons when it come up.

'You'd gotta work hides round on the west side. Course, out o' Low'stoft you dint do
that right a lot. You'd hev pieces o' old net sewn along the bottom o' the cod end instead.
Round the west side, though, you'd gotta hev hides. Yeah, a net wun't last very long
without 'em. They were bullock hides or cow hides what yuh used, an' they were just
lashed on the bottom o' the cod end more or less t' keep that from wearin' as it dragged
along the bottom. Thass surprisin' how they used t' save a trawl, yuh know. Yeah, they
used t' last quite a long while. Sometimes you'd git 'em when they were nice an' dry, but
other times you'd git 'em when they were right wet. You know, they'd just bin skinned.
When they were dry, they'd be right stiff, but as soon as yuh got 'em in the water they
went sorft agin.

'There used t' be quite a bit o' mendin' when you were workin' out o' Fleetwood. If the
net wuz too bad, you'd simply cut it orf, throw it overboard an' put a new one on. But if
you could mend it, then you would do. You used t' take a fair bit o' gear wi' yuh, but we
never used t' put anything ashore. No, not like they used to years ago. Well, that wun't
worth it, yuh see; you could allus buy what yuh wanted at either Milford or Fleetwood.
You'd carry two spare doors on board with yuh. Yeah, you'd rest them on the side o' the
casin', but all yuh other stuff would be down below – bobbins, bridles, floats, spare
trawls an' all that. You used t' carry two lots o' bobbins, big ones an' small ones, an yuh
bridles used t' be more or less the same. They'd be long ones an' short ones. Yeah, you'd
hev the long ones f' cod an' rooker an' the swimmin' fish, an' you'd hev the short ones f'
the soles an' plaice. You used t' swap bridles just like yuh did ground-ropes. That wuz one
o' the good things about trawlin' – you could change yuh gear t' suit yuh fishin'.

'One thing you dint carry wuz extra warp. No. All the warp yuh hed wuz on the winch.
Well, you couldn't very well carry a lot o' that. when you went round on that Westward
voyage, you'd praps hev a set o' warps on afore yuh went an' they'd last yuh out. They
lasted a good while, yuh know. There wuz some blinkin' strength in them 'cause they were
all steel wire. They dint rust all that much either really, but the strands used t' break an'

that'd make 'em a bit frazzled in places. If there wuz just one or two strands broke, you'd splice them up. But when they got too bad an' started t' frazzle out a lot, that wuz time t' renew yuh warp. Course, you could allus use an old warp. You could cut 'em down inta bridles an' all sorts o' things. Oh, there wun't nothin' wasted.

'There used t' be some blinkin' strain on the warps when the trawl wuz down an' you were in deep water. Oh yis. I mean, when yuh used t' come fast in deep water, the boat'd heave right over. You'd hetta stop straight away then. If the warps were too tight on a fastener, you'd go astern; but if you could heave up on the winch an' free it that way, you wun't bother. I remember once orf Padstow durin' the war, we hetta go astern then when we come fast. I dun't know what the devil we got hung up on, but we were messin' around f' nearly 12 hours an' we couldn't budge it. Yeah, we were 12 hours fast there in Padstow Bay. What the devil we got, I dun't know, but we managed t' git orf in the end when the tide turned. The warps slacked up a bit then, but we still couldn't haul it in prop'ly. We hetta cut one warp an' heave up on the other. When the trawl come up, the net wuz all in ribbons, so I should say we got a wreck o' some description. Yeah, there wuz one time there when we thought we were never goin' t' git clear of it. An' a day or two afterwards, blow me, if one o' the "Ocean" boats dint go an' catch the same thing. He lorst the lot, he did. Both his warps parted with the strain.

'That sort o' thing wuz why you always hed a man on deck watchin' the warps. See, so soon as ever you come fast, he could sing out an' the mate or third hand could go t' the winch an' start heavin' up. You wun't allus call the skipper out 'cause there wun't no need a lot o' the time. A little chuck-up on the winch would praps git yuh clear. What you mustn't do wuz keep goin' ahid, 'cause that'd pull yuh on tighter so you'd rip yuh net t' bits. A lot o' the time, when yuh come fast, there used t' be one warp a-dudderin'. Yeah, that'd tighten up an' keep a-shakin'. When that did that, you knew you were somewhere you dint oughta be. Course, when you were workin' rough ground, yuh warps used t' dudder then, but that wuz a different kind o' thing alt'gether an' you got t' know that. Yis, that wuz different t' when yuh come fast.

'Round Padstow we used t' work a dan. Yeah, soon as ever we got a nice bit o' ground, we'd put a dan down. We did that when I wuz in the "Tritonia" an' all. Oh yeah. I mean, anywhere where you were on a nice bit o' fish, you'd put a dan down. You hed it all riddy; you only hetta drop it over the side. Course, in the steamboats, when you were goin' afore the tide, the skipper'd tell yuh t' ease down two or three revs, or praps three or four. But when you were comin' back up the other way, he'd want them back. Yeah, when you were stemmin' the tide, he'd want them revs back. When yuh go with the tide, thass surprisin' how far yuh go, yuh know. Yeah, an' then you'd gotta come round an' git back up agin t' where yuh started. That'd take yuh twice as long some times gittin' back as it did goin' down.

'I used t' be down below when they shot the trawl. I dint hev nothin' t' do wi' that business. Soon as ever the trawl wuz over, the skipper used t' blow down the tube — you know, give yuh a shout — an' ask f' so many revs. An' that wuz that. Us engineers used t' take our watches independent. Oh yeah. One of us would take from breakfast t' dinner t' start with. That wuz eight o' clock breakfast, but dinnertime all depended on what time they hauled. If that wuz a long tow, you might hev dinner afore yuh hauled; but if that wuz a short un, you'd probably hev it after. After dinner one of us used t' relieve the other an' go on watch till tea, then you'd change agin an' run through t' midnight. From midnight you'd go t' breakfast an', oh, that mornin' watch used t' be a long un! Still, that wuz better'n the crew's watches. Oh yis, definitely. Yis, I mean, the crew hetta be turnin' out praps every two or three hours if you were on a good fishin', but us engineers used t' do our watch an' then turn in. Oh, we used t' git a good ow sleep, we did.

'You used t' hetta watch yuhself on board, yuh know, 'cause you could soon hev an accident if you wun't careful. When you were haulin', you hetta be very careful when you were hookin' the door up inta the gallus. See, the man at the winch, he might drop the door afore you were riddy an' you were liable t' git yuh hands caught in the sheave. You hetta make sure an' sing out good an' loud t' let him know you were riddy f' him t' drop the doors. One boat I wuz in, we hed a young chap who got his arm caught through the sheave when the bloke on the winch dropped the door too soon. Cor, he wuz in a bad way, he wuz! We hetta come straight inta port with him. An' that finished him. He dint go t' sea no more.

'Another thing what could happen wuz yuh bag-stopper might break. That wuz a rope what run from yuh forem'st riggin' down t' near the winch. When the cod end come in, that used t' swing aginst that an' stop. Well, if that brooke, an' I hev sin it, the cod end would swing right acrorss the deck an' maybe give someone a clout. An' sometimes, the man whass on the gilson, praps he'd heave up too high an' the cod end would go right over the top o' the bag-rope. Oh, yeah, thass orften bin done, that hev. Once, on the *Tritonia* there, we got s' much fish one day that the weight brooke the forem'st in three places! Yeah, our mast come down in three pieces wi' the weight o' fish. I only saw that happen the once. We hetta go inta Fleetwood an' hev a new mast put in. We hed a lot o' cod an' whitin' in that partic'lar bag, I remember, an' I should say that the reason the mast come down wuz that the forestay musta brooke first. You know, the wire support what run right onta the bow. Thass the main support, that is, an' I should say that musta brooke just as the bag swung in. I can't see a mast comin' down like that otherwise. An' that wun't an old mast either. I mean t' say, this wuz just after the war an' all the boats what'd bin under the Admiralty hed a complete refit afore they went back t' fishin'.

'When we were out here in the North Sea, we never used t' go very far. Well, you hetta watch yuh coal, yuh see; you dint carry all that much, not like the steam trawlers did. we used t' git as far as the Leman Bank or out t' The Knoll, or praps we'd run up southerly t' the Hinder. After you'd bin herrin' catchin' on the hoom fishin', you'd git praps a couple o' trawlin' trips in afore Chris'mas, then praps three or four afterwards, an' then you'd be orf t' Fleetwood. The furthest we used t' run from here wuz acrorss t' the Brown Ridges. You used t' git some nice plaice acrorss there. Well, so yuh did at The Knoll. There used t' be some good fishin' up southerly an' all. See, the big trawlers dint bother very much with the Galloper an Hinder an' them places, so you could always drop in there an' fish a trip or two.

'We used t' land in here in the Waveney Dock. See, the herrin season wuz over when we used t' go trawlin', so we could git in there all right. The lumpers used t' land yuh fish. You never used t' do that, not after the war. Durin' the war you did, but not when you come back t' Low'stoft afterwards. When I wuz round Milford in the *Constant Star,* you hed the sailors come aboard an' help yuh land. Yis, four or five o' them used t' do that t' earn a bob or two. Course, at that time you used t' shelf quite a lot o' fish. Yuh long stuff, you did, cod an' that. The mate wuz in charge o' icin' the fish, an' there wuz an art in that. Oh yis, you can soon spoil a trip o' fish if you aren't careful. That you can. The long stuff hed gotta lay light, specially cod an' haddicks; but the flatfish, you could stick them in on top o' each other.

'You git a lot o' hake round the Westward. Or you did do. I think thass a game practically finished now, hake fishin'. Well, you dun't hear all that much about it, anyway. Durin' the war you used t' git loads. You used t' git whole blinkin' trips on hake sometimes. But then, the latter part o' the war an' just after, the foreigners started comin' round pair-fishin'. Thass where two boats tow one trawl betwin' 'em. Well, they used t' just sweep 'em up doin' that. Thass a blinkin' rum width, that net is, with one trawl warp on each boat. That cover a big area. No wonder there ent no fish about when yuh catch 'em like that! That musta cleaned 'em out. When I wuz round at Milford, you only hetta go a little way out afore you'd put yuh trawl down an' drop onta some hake.

'There used t' be a good bit o' turbot an' brill round the west side as well. You used t' git them in the bays a good bit. So yuh did lobsters an' crayfish. Cor, you used t' git baskets full o' them! We used t' keep quite a lot of 'em for ourselves an' the rest we used t' sell. Yeah, there wuz always a market for them. The Navy blokes used t' know all about 'em, I can tell yuh. Yeah, they'd soon come alongside arter a basketful. Course, we got plenty o' cigarettes in return, you know, or a bottle o' whisky. Oh, they used t' treat us well − specially the little patrol boats. They used t' come right up alongside yuh for a bit o' fish, so we were never short o' cigarettes when we were round there durin' the war. No, we dint mind givin' away a basket or two o' fish f' cigarettes!

'Course, fishermen hev a ration o' fish now an' they hefta keep to it, but when I wuz in the *Tritonia* you could hev what yuh liked. Yeah, the mate used t' dish 'em out, an' you might finish up with as many as you could carry. There's a blinkin' difference now, ent there? I mean, we used t' be allowed soles. Only skippers an' mates can hev them now. An' what do they fetch on the market? − £150 a kit or more! Thass money, ent it? The only stockie bait we used t' git on the *Constant Star* wuz shellfish. Yeah, we'd hev the queens an' that sort o' thing − you know, the scollops. They used t' sell round on that west side all right. Oh, they liked them round there. When I wuz in the *Tritonia,* the only stockie we used t' git there wuz rolls − cod rolls mostly. But that dint matter what stockie yuh got when I wuz at sea, you were never restricted on yuh allowance o' fish. No, you could hev what yuh wanted more or less, where t'day − well, fish is like blinkin' gold dust, ent it?

Mid-water diesel trawler 1955
70 tons net.

CHAPTER SEVEN

The Coal Gobblers
Steam Trawler Reminiscences

'Don't call me a common fisherman any more.
Don't call me a common fisherman any more.
Fresh fish to you I bring;
Don't call me a common thing.
I'm as good as them that work upon the shore!'
(Traditional fishermen's chorus)

The major trawling ports of the great age of steam fishing were Hull, Grimsby, Fleetwood, Milford Haven, Aberdeen and North Shields which developed after 1880. There was no sizeable fleet of steam trawlers at Lowestoft until the 1920s and 30s. Prior to that, as we have seen, trawling in the Suffolk port was carried on by sailing smacks working the grounds near home and its migratory drifter-trawlers. Nevertheless, trawling became a popular occupation among working Lowestoftians because it had two distinct advantages over herring-catching. Firstly, crew members (except the skipper and mate) were paid a regular weekly wage instead of drawing a share of the boat's net earnings at the end of a voyage – remuneration that might be very small or even non-existent in the event of a bad voyage (see 'The Driftermen', chapter six). Secondly, the work was less seasonal so there wasn't as much chance of a boat being laid up at slack periods.

This latter held generally true even during the depression, when bad markets and rising costs often did result in unsuccessful steam trawlers being taken out of commission for a trip or two. Even if this happened, however, the trawlerman still had one great advantage over his counterpart on the herring drifters: he received unemployment pay. The drifterman didn't get dole because his share payment status classified him as self-employed, and this meant that when he was out of work for any length of time he had to claim poor-relief, with the indignity of food tickets and the means test. Its arbitrary and unfeeling administration, still remembered with great bitterness, caused many Lowestoft fishermen to feel that trawling was a better way to earn a living than drifting.

Fully fledged steam trawlers (as opposed to drifter-trawlers) did not appear in Lowestoft in any numbers till after 1918 when several local fishing companies began to invest in them as a hedge against the uncertainties of the herring trade. There was a general mood of optimism prevailing in the fishing industry immediately after the war. People believed that things would be just the same as in the peak years up to 1913. But this was not to be. Catches dwindled as stocks were depleted by the increased efficiency of boats and gear, and because of the economic depression there wasn't the buying power at the bottom end of the market. Halibut, turbot and soles sold all right, but the demand for cod and haddock was very uncertain.

Still, this wasn't immediately apparent at the return of peace and steam trawlers were purchased to fish out of Lowestoft. Most of them were vessels that had been on Admiralty charter during the war and they now came back to their normal way of life after extensive reconditioning and refitting. Several of them were ex-Aberdeen boats, with a few from

The launching of the drifter trawler *Merbreeze* in 1931, the last coal burning boat to be built for the Lowestoft fishing fleet.
Below: the steam trawler *Amethyst H455,* 146 tons, 96 h.p. coming alongside in Hull. She was built in 1928 for the Kingston Steam Trawling Co. Ltd.

Grimsby, and they were expensive to buy and to maintain. Herring fishing involved a good deal of capital expenditure to purchase, equip and run boats successfully; trawling involved even more. For one thing, steam trawlers were bigger than drifters and burnt a lot more coal. Trawling also caused much more wear and tear on engines and gear than drifting. In short they were much more expensive to operate, especially as they were driven further and further afield in search of good chatches. For these reasons nearly all the Lowestoft companies that bought steam trawlers in 1919 and 1920 had got rid of them two or three years later.

The vessels were all around 110 feet in length, of 75 tons net and with triple expansion engines of 67 HP or so. They had been built, for the most part, during 1910-11 and were intended for fishing well out into the northern expanses of the North Sea in deepish water. This limited their success when working out of Lowestoft in much shallower conditions, a fact acknowledged in the remark that one always hears about them: 'They were too big for the port'. One thing is certain — the companies that bought them lost a good deal of money.

It wasn't just a matter of low earnings either. The primary loss was incurred through depreciation, which was very high on steam trawlers at that time. Just consider a few examples. The Resolute Fishing Company purchased three steam trawlers in 1919: the *Loch Broom LT327*, the *Strathlossie LT511* and the *Loch Eriboll LT707* for £9,050, £9,750 and £16,700 respectively. After a very poor performance the *Loch Eriboll* was sold in 1921 for £7,250, and the *Loch Broom* and *Strathlossie* followed two years later for £3,250 and £3,050 after a year or more of barely covering running expenses. The firm of Jack Breach Ltd. had two steam trawlers, the *Strathfinella LT531*, bought in 1919 for £10,750 and the *Ben Iver LT788*, which cost £17,250 one year later. The *Strathfinella* was pretty successful in 1920 and 1921, but thereafter her earnings declined considerably and she was sold in 1924 for £3,000. Her companion had a reasonable year in 1921, but failed to make much impact afterwards and was put on the market for £5,000 in 1923.

It is this kind of arithmetic that sums up the story of steam trawling in Lowestoft in the early 1920s. And yet, just as the local companies were selling off their boats, a firm from outside the town began to build up the biggest fleet of fishing boats of any kind ever based in the port up till then. The Consolidated Steam Fishing and Ice Company Ltd. of Grimsby opened up a shipwrights business in Lowestoft just after World War One, which was obviously a springboard for the main enterprise, because in 1923 they began to bring steam trawlers into the port. By 1930 they had around 30 vessels based at Lowestoft, all of them under the managership of George Frusher, a man who had been a Consolidated Company skipper in the days of the Dogger Bank fleets and who had also commanded a steam-cutter that ran the fish from the trawlers down to Billingsgate.

The Crown Boats, as they were known, because of the company's house-flag painted on the funnel, were a notable feature of the Lowestoft scene between the wars. All of them were superannuated affairs, which had been built in the 1890s for fishing the Dogger Bank and beyond, and which had now really come to the end of their useful working lives. Rather than sell such boats on a depressed market, the Consolidated Company moved them en masse to Lowestoft, hoping that they could pay their way by fishing the nearer grounds of the southern North Sea. Some of them did run across to Borkum and

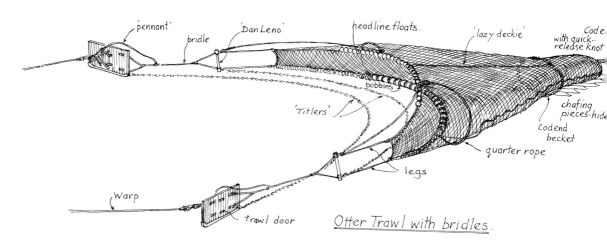

Otter Trawl with bridles.

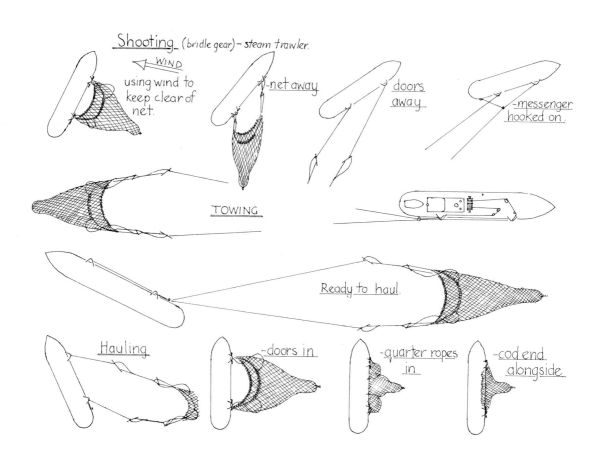

Shooting (bridle gear) – steam trawler.

using wind to keep clear of net.

net away

doors away

messenger hooked on.

TOWING

Ready to haul.

Hauling — doors in — quarter ropes in — cod end alongside

Heligoland at times, but for the most part they worked closer to Lowestoft than they had done to Grimsby, thereby cutting down on running expenses. Another factor that influenced the shifting of so many boats southwards was the over-fishing of the Dogger Bank area, which was no longer able to support the intensive trawling it once had. Later the Consolidated Company invested in new distant-water trawlers for Iceland and the Barents Sea, as well as modernising the boats of her associated companies in Cardiff and Swansea. All of this cost a lot of money; and if it so heppened that old trawlers could remain profitable by working different grounds, then the firm was happy. If they didn't pay their way, then they were laid up for a trip or two and the crews stood off.

The steam trawlers worked the conventional otter trawl, but during the early 1920s a refinement of the method was introduced from France and pioneered by boats belonging to the firm of Neale and West in Cardiff. What it did was to locate the otter door further from the net by introducing a single wire bridle (of varying length, according to what kind of ground was being worked and what species of fish sought) between them. This Vigneron-Dahl gear, as it was known, increased the size of trawls by virtue of the fact that the bridles enabled the mouth of the net to be spread wider. The headline and ground-rope were kept apart at either end of the net by iron-shod wooden posts known as guindineaux. The term soon became changed to Dan Lenos!

The use of this gear necessitated the payment of a royalty to the French government, and it wasn't long before the fishing companies were trying to find a way round this. In the end a simple adaptation proved good enough to beat the patent rights. The original Vigneron-Dahl gear had a hole in the otter door, through which the bridle ran on to the winch once the door was chained up to the gallus. Someone came up with the idea of having a wire strop, independent of both warp and bridle (hence it became known as a pennant), but joined to them, which could act as a link between the two and bypass the door altogether once the latter was secured to the gallus during hauling. While fishing was in progress, it simply remained suspended between the bridle and the warp, clear of the door so as to prevent chafing. Once the door was chained to the gallus head, though, both the warp and the pennant were unclipped from the door's bracket and allowed to run through on to the winch, drawing the bridle after them until the net came up to the side of the boat.

George Stock had many years at sea on the steam trawlers of this period, working otter trawls of all dimensions, and types:

'I went inta trawlin' because driftin' wuz more of a haphazard business. You got a weekly wage in a trawler, but in a drifter you were on a share. You couldn't go on the Labour if you were driftin', yuh know, but if you were out o' work from trawlin', you could. Oh, yeah, trawlin' wuz more secure. That wuz a better deal. A lot o' the Low'stoft men couldn't stand t' be in a drifter all year 'cause they'd likely git stood orf. But a lot o' the boys from out o' the country villages went driftin' – specially hoom fishin' time. Now I dint like the sea all that much, but I pushed m'self t' git on. I got my tickets an' worked me way up. I used t' hate goin' away from hoom, an' yit every trip is a voyage in its way. Yeah, an' then there wuz the excitement when the ow cod end come up an' you see whass there.

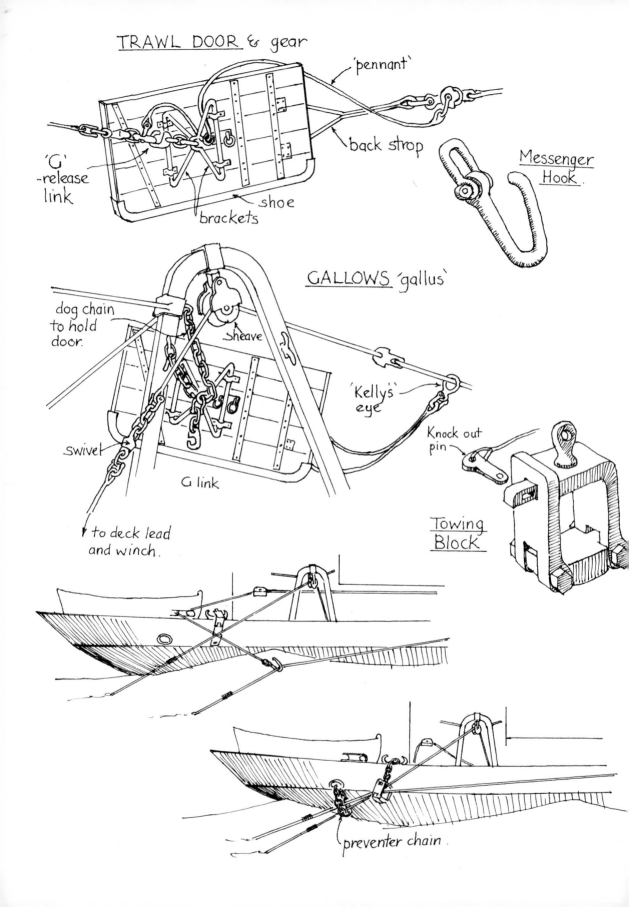

TRAWL DOOR & gear

'pennant'

'G' -release link

back strop

shoe brackets

Messenger Hook.

GALLOWS 'gallus'

dog chain to hold door.

sheave

swivel

G link

'Kelly's' eye

Knock out pin

Towing Block

to deck lead and winch.

preventer chain.

THE COAL GOBBLERS

'I wuz in the ow Crown boats f' quite a long while. I musta gone in about 13 or 14 of 'em alt'gether. Every time one of 'em used t' go wrong – you know, hev the boiler tubes pack up or suffin' – they just used t' go an' git one out o' the creek, where they used t' lay 'em up, an' send it out fishin'. Cor, there dint half use t' be some rats aboard them! They were full o' rats. Yit they wun't touch the cheese what we hed t' eat. They used t' put these Dutch cheeses aboard an' they were so hard that the rats wun't even touch 'em. When I wuz mate along o' Albert Lockwood, he used t' cut these cheeses up an he used t' reckon you could sole yuh boots with 'em. As the boys were guttin' up, he'd chuck bits o' cheese at 'em for a laugh. I spose the Consolidated Comp'ny hed about 30 or 40 boats in Low'stoft alt'gether. A lot of 'em were the old aft-siders – you know, with the wheelhouse aft-side o' the funnel.

'We used t' make big iron kettle full o' cocoa f' the night on board the ow Crown boats, an' when you lifted the lid orf there used t' be all grease floatin' on top. That wuz made specially f' trawlermen, this here cocoa, an' that wuz just like brick dust. That'd stick in yuh throat when you drunk it. You just used t' keep warmin' it up as yuh used it an' keep puttin' more an' more water in. Even t'day I can't stand cocoa! The tea we hed wuz rum stuff an all. You used t' find no end o' nails in it. They'd bin put in t' make the weight up. The comp'ny used t' supply it, but I dun't know where they got it from. That used t' come in gret big boxes. I've told yuh about the cheeses alriddy – how the rats wun't even touch 'em! Then there wuz the meat. That used t' be contracted aboard from the butcher at about 4d a pound. We used t' laugh an' say, "They mighta left the hoofs on. We'd hev the lot then!" Cor, that wuz rough stuff! But you survived.

'Thass surprisin' how much them ow boats'd stand. I wuz mate in the *Volta LT180* along o' Jack Hunter an' we were comin' up over the shoals inta Low'stoft. That wuz blowin' hard an' she took this here sea. All the wheelhouse winders come smashin' in. "Shall I ease her down?" I say. "No! Keep the bugger goin'!" he say. "Dun't ease her down." We were steamin' hoom t' catch the mornin' market, yuh see. Well, coo t' heck, if we dint hefta keep our hids down all the way in, otherwise we'd ha' got blinded by the spray. About a couple o' years after that, I went skipper o' that boat. Cor, she wuz an antique, built about 1890. Well, they all were. An' rats! Rats forever!

'If you brooke down at sea, you used t' run a sack up yuh forem'st. Someone'd see yuh an' say, "Cor, there's a hovel!" An' they used t' come an' tow yuh back inta port. Then the companies would hefta agree how much t' pay for the tow in. Any boat towin' a Crownie in wun'ta got much, I shouldn't think, 'cause none o' them ow trawlers were worth anything. Course, you used t' git a lot o' boats wantin' help when the weather wuz really bad. I remember when I wuz mate in the *Brent LT98* an' there wuz a bad blow one hoom fishin' time. That'd be about 1928. An' that blew – oh, that did blow! We were down from Low'stoft, along shore, so we went out easterly t' git away from the shoals. we ran up south an' then come in, an' we got a trip o' dogs an' rooker just t' the back o' The Barnard. I know we made £200 odd. That wuz the top trip by far. Some of 'em dint hev above 15 kit f' the whole trip! They just hent bin able t' fish. My skipper wuz ow Billy Chilvers, an' all the way comin' in we could see these drifters what'd bin parted from their nets. Then there were the smacks, with the masts all gone an' that. They all wanted help, but you couldn't tow 'em all in.

'Course, there wun't no radios in them days — well, not many anyway. The Crown boats dint hev 'em, you can bet! Mind yuh, when I wuz in the *Castlebay LT1295* wi' my brother, we did hev a radio in her. That wuz one o' the first receivin' sets 'cause I remember John Utting, the owner, sayin' to us, "I'm goin' t' buy you crew somethin' special when you go away t' Padstow." This'd be about 1924. That wuz a funny thing, that wireless. A sort of a crystal set thing. An' ow Sam horn, what wuz mate, he used t' go up t' the set an' swear like mad if a foreigner come on the air. "Speak bloody English!" he used t' say.

'Like I say, the Crown boats dint hev radios. Ow George Frusher got onta me there once f' bringin' up at Southwold. That wuz gittin' towards Chris'mas an' that wuz a-snowin' an' a-blowin' like the dickens. Boats were a-dodgin', so we went an' dropped our hook just orf Southwold Light. That wuz a-blowin' from the West'ard an' our trawl wuz split, so we lay there an' got mended up f' the next trip. When we got inta Low'stoft, ow Frusher knew we'd brought up. He started goin' on about when he wuz on the Billingsgate cutters. "We dint stop f' nothin'," he say. "I never brought up, in any sort o' weather." I say, "No, but you wunt in the sort o' ow carts what we are!" I mean, Low'stoft could be a bad harbour sometimes — specially when you got an easterly wind, or a sou'easterly, an' a big tide. You'd be comin' in wi' the swell up behind yuh an the tide sweepin' acrorss, an' that could be a bit nasty.

'Durin' the 1920s and 30s the North Sea wuz gittin' fished out. That wuz why so many o' the Low'stoft drifter-trawlers went round t' Milford Haven an' Fleetwood. There musta bin about 100 local boats round there alt'gether out o' Low'stoft an' Yarmouth, an' a lot of 'em would stop there all the time. On top o' that, the fish trade itself wuz in a bad way. Bad markets an' all the rest of it. Blimey, you could buy a good trawler about 1928, 1930, time f' under a thousand quid! I think the *Junco H587* corst only about £800 when she wuz sold — an' she wuz one o' the newer ones. You could git an ow fleeter, one o' Hellyer's or the Gamecock boats, f' about £300. They were the ones what'd bin built about 1903 or 4. Another thing at that time wuz yuh money. If you were a skipper or mate, you used t' square up every three trips; but if you'd done badly, you praps wun't hev anything t' come. An' there wuz no goin' on the labour if you got stood orf. Yuh wife used t' draw a weekly allotment time you were away an' that'd be deducted from yuh share when yuh settled.

'Skippers used t' git a few little perks o' course. The coal wuz one. Yeah, the skipper got a commission orf the firm he ordered his coal from. So he did orf the grocer sometimes, if he wuz lucky, or orf the butcher. They dint all give it, but you'd praps git a Chris'mas box from 'em if you were a good customer. By the time I went skipper that wuz mostly a case o' dealin' orf who the comp'ny told yuh to. You hed yuh orders where t' go an' you obeyed 'em. Even if you wanted t' shift, you couldn't. Skippers' an' mates' food bills used t' be deducted from their share. Thass why some of 'em used t' be so tight about grub. The comp'ny used t' divide the food bill f' the trip by the number in the crew; then they'd knock orf the skipper an' mate's amount from what they'd earnt. Well, you can see why they dint want the crew t' eat too much. An extra pot o' jam used t' be a crime!

'Now I wun't like that. I always reckoned that if you fed yuh crew well, then they'd work for yuh. I remember comin' hoom from Fleetwood once, when I wuz in the *Amalia*

LT241, an' Charlie Dance, one o' the Vigilant Company's bosses wanted t' know why my food bill wuz two or three quid dearer than some o' the other boats. So I told him. An' ow Fred Spashett, who wuz the main boss, backed me up. He say, "You're the skipper. You're in charge. You carry on as you think right." Jack Breach told me exac'ly the same thing once round at Milford. "George," he say, "You're right. Treat the crew good an' let 'em feed well — they work a lot better for it." An' that wuz so. They did do.

'When I wuz in the ow Crown boats we used t' hev a cask o' brine aft-side o' the wheelhouse t' put bits o' beef in f' saltin'. We used t' put saltpetre in as well an' you knew that wuz riddy f' usin' when you could float a potato the size of an egg on it. The cask used t' be lashed in between the door t' the galley an' the mizzenm'st. There used t' be a little space there where you could stand it. Course, the ow iron mizzens they hed on them boats used t' act as a funnel f' the cabin fire, an' some of 'em did f' the galley as well. The masts were hollow an' there were slits in 'em three-parts o' the way up t' let the smoke out. When they rusted through, they hed wooden masts put in an' hed proper funnels f' the cabin an' the galley. I wuz in the *Aberdeen LT123* an' the *Scarborough LT136,* an' they were boats like that. Yeah, an' the ow salt beef cask wuz always there. The joints wun't too bad t' eat after they'd bin brined. That seemed t' take some o' the toughness out of it an' that used t' hev a lovely glisten after you'd boiled it.

'Course, bad grub wun't the only thing on board some o' the boats. Some o' the ow skippers used t' be devils an' all. Specially when that come t' dishin' out the stockie. If you dint watch 'em, they'd be inta the pub an' drink it all away. I wuz with one or two like that. They'd come out o' the pub about three o' clock in the afternoon an' give yuh praps a shillin' out o' what wuz left. That'd bin drinks all round f' them an' their friends. An' even if there wuz any left, you wun't see it. Some of us went t' the Custom House about it once, but they told us there, "You can't do nothin' about it. As long as he give yuh suffin, he's within the law." Well, that wun't right, wuz it? At one time o' day the crew used t' git all the small rooker an' gurnets as their stockie, but then that got so the companies used t' pay the blokes about a pound each every trip. When I wuz skipper I always made it a practice t' send one o' the crew t' the comp'ny orffice t' git the envelope wi' the stockie in it, an' he'd bring that aboard unopened. Some o' these ow skippers used t' say, "You're daft! You're wet!" But I knew I wun't. You treat yuh crew well an' they do the same f' you. An', anyway, that wuz their money as well as yours. They'd earnt it.

'Before they started payin' yuh stockie direct to yuh in money, all the fish you were allowed f' stockie used t' be in heaps along the front o' the market where the water pumps used t' be. You know, where the posts came down what used t' hold the roof o' the market up. The main catch, yuh plaice an' cod an' stuff, stood further back. They used t' start orf by sellin' the main trip, then a junior salesman would come along an' sell orf the stockie fish. That used t' be bought mainly by the little fish friers an' hawkers. They used t' come down wi' their barrows an' you'd see 'em loadin' up in any open space they could git. After the Second War, when the diesel boats come in, a lot o' the comp'nies did away wi' stockie an' put everyone on a poundage.

'After yuh fish wuz sold, the mate used t' take the boat up-through-bridge t' coal. There used t' be all the railway trucks drew up there along Commercial Road an' that wuz where all the coal wuz. For the trawlers anyway. The smacks an' the drifters used t' coal at the market orf the ow coal wherries. I remember when I wuz deckie in the *Strathfinella,* we used t' land early if we could. Then we'd go up-through-bridge t' take on coal an' ice. You wun't allowed t' go t' the Ice Comp'ny wharf because o' the fresh water pipes they hed there. They were frightened you might break 'em, so the ice wherries used t' come up alongside yuh. We used t' be there from about ten o' clock in the mornin' till three in the afternoon. Then we'd be away t' sea agin the next mornin'. They wun't do that t'day, would they? But you'd gotta do it then. You dint hev no option.

'An' not only that − I never could see the harm in hard work. That wuz all in a day's carry-on, if yuh see what I mean. Thass like when I wuz skipper durin' the war. They sent me a big ow Frenchman, what'd got out o' France away from the Germans. They shipped him aboard as trimmer. Well, he couldn't git down the little round bunker holes in the *E.W.B. LT1124,* so I trimmed. That wuz either that or go back an' lay till they got someone. I dint mind. I'd done trimmin' in the *Strathfinella.* An' anyway, a real interested chap, he should be able t' turn his hand t' anything.

'Out here in the North Sea you used t' start orf here in January goin' up-along, as they used t' call it. That'd be south-about, up t' The Goodwins an' The Falls. Well, you'd work The Shipwash an' The Gabbards as well. You used t' leave The Knoll in the early part o' the year 'cause all the plaice hed shot their rows then an' were nothin' but skin an' bone − slabs, as we call 'em. We used t' go up along when I wuz in the *Brent* along o' ow Billy Chilvers. He wuz a real Knoll man, he wuz, but that wun't no good stickin' there once the slabs set in. We used t' go up as far as the London River an' that wuz the ordinary otter gear he used t' like. He used work bobbins as well, the wooden ones, made out o' ellum or ash, an' maybe a titler. O' course, the French bridle gear became very popular about 1927, 1928, but ow Billy never would shoot it. He'd bin a smack skipper f' years afore he went inta steam trawlers, an' as soon as we got out he'd say, "George, git our gear rigged up." So what we used t' do wuz take orf the bridles an' that what the firm'd give us an' git the ow on-the-door gear fixed up.

'Now as regards bridles, I always found 'em a good thing − especially where you were after long stuff. You know, the swimmin' fish, like cod, hake, whitins an' haddick. They were a big advantage then. Yuh trawl dint bite inta the ground like the ordinary otter gear. I've worked ten foot bridles, 15 foot, 30 foot an' 40 foot. I've even bin up as far as 75. That wuz the longest I ever worked. Some o' the big Iceland boats used t' go up to about 150. Mind yuh, we done well on the ow *Brent,* usin' on-the-door gear. She wuz only a little ow trawler, an' she could only stop out about eight days 'cause she couldn't carry much coal, but she fished well. On the way hoom ow Billy used t' say t' me, "George, hev we got plenty o' stockie?" An' if we hent, we'd come inta the Thornton Ridge an' git a few good hauls o' small rooker an' other stuff.

'You used t' git some turbot an' brill goin' up-along. Yes, that yuh did, an' you'd git 'em daytime as well as in the dark. We used t' go up as far as the Varne Bank, orf the French Coast, an' I remember gittin' one or two nice trips there o' turbot an' blond rooker. We used t' go in orf Dover as well. There wuz a gulleyway there an' you could drop onta soles. The only trouble wuz that the ow Crown boats dint really hev enough power t' tow aginst the tide up there, so you could only work goin' with it. You hetta give it up a lot o' the time an' come back down inta the North Sea, where there wun't s' much tide. I remember bein' up there one year, when I wuz skipper o' the *Elector LT579*. Well, she hed only a very little power an' she just wun't pull a lot o' the time. Now she wun't a Crown boat; she belonged t' Arthur Gouldby, but she wuz one o' them ow North Sea trawlers.

'In the spring you'd be anywhere down along the shoals, out orf Cromer. You could git some good trips there. So yuh could on the Indefatigable Bank. Yeah, you could drop onta some nice plaice there, an' turbot an' brill. The Brown Ridges were good grounds too, but I never saw any point in gittin' over t' Skillen an' them places 'cause you used t' git s' many o' them little ow plaice. Ivy leaves we used t' call 'em, or penny stamps. You wun't want a small mesh for 'em or anything – you just got 'em! An' you couldn't always sell 'em when yuh got 'em hoom. I've sin kits an' kits o' them standin' around on the market. Some o' the Consolidated skippers, like Charlie Larter an' Mad Jack Braddick, they continually worked well over easterly an' they always used t' git a lot o' these small plaice. About the only people what'd buy 'em wuz the fish friers. They just used t' cut the hids orf an' fry 'em whole, on the bone. There wun't much on 'em, but they were lovely an' sweet.

'Another thing about workin' over Skillen an' Borkum way, an' down t' Heligoland, wuz the distance there an' back. That wun't s' bad f' the big boats like the *Valentia LT150* an' the *Valeria LT156* 'cause they could carry plenty o' coal. But with the smaller ships, you'd gotta weigh up how far yuh coal would go. You know, if yuh got a westerly wind, an' that come on strong, you'd be punchin' all the way back – an' away go yuh coal. You always wanted t' allow at least two days coal f' that run an' you'd always say t' the ow chief, "Let us know how the coal is." Another thing wuz the weather. If yuh got a strong westerly blow over there, you could cop out. I remember Jack Braddick comin' in once with the wheelhouse washed away after the boat'd shipped a sea.

'Mind yuh, the further north yuh went down, the better quality the fish. Yeah, you could git some nice trips down the Amrum Bank an' Heligoland. That wuz a better class o' fish alt'gether. Big plaice. An' good haddicks as well at one time. We worked there one year when I wuz mate along o' Albert Lockwood in the *Exeter LT139* an' we did out an' out well. That wuz March time, an' thass them times when yuh got good hauls what stick in yuh mind. You dun't wanta remember the bad! When yuh did well, you'd praps write it down in yuh book. You know, where yuh were, what yuh got, how long yuh towed an' all that. Oh, that paid yuh t' keep a record. You could always look back then an' see what you'd done the year before.

'We used t' git down on the Dorgger Bank o' course. Yeah, I've worked down there nearly all times o' the year. You used t' catch tons o' haddick there at one time, specially in the Botany Gut, but then them Dannish seine-netters come an' cleaned it all out. Cor, I've sworn about them! Another good place used t' be the Silver Pits. See, that wuz another hole – an' where yuh git holes yuh git fish. There wuz another place down there where we used t' come orf the edge o' the bank an' git inta what they called the amber weed. You'd be comin' orf the Dorgger just t' the nor'ad o' the South Rough an' you used t' git some nice trips o' plaice there. One or two o' the skippers used t' work the South Rough all the year round. Jack Hunter did. When I wuz wi' him in the *Fidelia LT187* we did pretty well down there. Mind yuh, you got plenty o' splits 'cause there wuz a lot o' boulders an' that on the bottom. That wuz why they called it the South Rough.

'One thing I found out durin' my time fishin' wuz that it always paid yuh t' keep yuh ears open. I wuz always listenin'. Thass how yuh learnt. When I wuz skipper o' the *Volta,* I went over t' Skillen there once 'cause I'd heard some o' the ow boys talkin' about a gulleyway near where the Germans sunk the *Cressey* an' them other two warships in the First War. Well, when we got there, there wuz no one about an' we got in four or five days good work. We were gittin' all sorts, an' when we come in the trip made £200. An' thass how yuh used t' go on, gittin' in where yuh could an' earnin' a livin'. That partic'lar trip we chucked a lot o' the small whitins back an' I spose the blokes what unloaded the fish musta heard the crew talkin' about it. When I went up t' see ow George Frusher, he say t' me, "I hear you're bin throwin' whitins away." I say, "Only the small ones. There wuz too many t' gut. That wuz freezin' weather an' the blokes wanted a nap. Anyway," I say, "we got plenty o' other stuff. We've made a good trip." He say, "That ent the point. You coulda hed more. I dint worry about people hevin' a nap when I wuz skipper." An' thass how he used t' carry on. But I knew different. You can't drive blokes for ever.

'When I wuz with the Consolidated, most o' the skippers were lone hands. You'd see other boats, but o' course you never hed no wireless t' talk t' people with. If yuh wanted t' speak t' each other, you hetta draw up alongside an' shout. Yeah, you tended t' work on yuh own, though a lot of 'em used t' git t'gether in the pubs. You know, the skippers. Yeah, they'd git in there an' swap talk. All o' them pubs along Commercial Road used t' be full o' skippers. Course, they're nearly all shut now, but thass where they used t' git. An' not just after a trip! A lot of 'em would git in there afore they sailed. That they would. There wuz several of 'em used t' go aboard blind drunk, an' the mate'd hefta take the ship till they sobered up. When I wuz round at Milford, there wuz an ow skipper in one o' the Ramsgate trawlers, the *Garrigill R267* or *Lady Luck R355,* one o' them boats, an' he always used t' go aboard drunk. They used t' bring him down in a taxi an' put him aboard an' away the boat used t' go.

'I never believed in that. You wun't in charge o' yuh boat when you were in that state. An' some o' them boats used t' take some handlin'. The *Amalia* did. I wuz in her about eight years an', oh, she wuz an awkward sea ship in some ways. Mind yuh, she hed a lovely big stern, so that if a sea wuz a-runnin' with yuh that'd push the stern up an' you could ride it. She wun't bad hid at it either. You know, when there wuz a blow, you could dodge her till the cows come hoom. That wuz broadside where she wuz bad. She used t' roll about s' much. All them Dutchmen did. There wuz about eight of 'em brought inta Low'stoft about 1925 or 6. We used t' reckon that the *Amalia* wuz give away in wi' the rest! Oh, she wuz a beggar broadside on. Now the ow *Rosalind LT977* wuz completely different. She wuz lovely. She wuz a daddy. One o' the nicest sea ships I wuz ever in. She wuz one o' Hellyer's ow fleeters. They named all their boats after Shakespeare's characters, yuh know, an' she wuz one. The *Blanche H928* wuz another. Ow Oscar Pipes hed her round at Milford. I spect you're heard about him. He wuz the one what used t' poach in Irish waters. Cor, he wuz a rum ow boy! He saw his own son git killed an' the blokes reckoned he dint turn a hair. The boat come fast an' a warp flew up an' nearly cut his hid orf.

'You used t' hefta watch the warps when yuh come fast 'cause first they'd drop, then they'd fly up. Sometimes they'd part. I've hed parted warps many a time. They'd fly about then, sometimes. If they parted down near the doors, that wuz all right. But if they parted at the towin' block, that wuz when they were dangerous. They used t' whip about then. Mind yuh, I never hed no real bad accidents through that, thank goodness! Another thing what could happen wuz blokes used t' git caught in the doors when you were haulin'. You used t' hook the door up inta the gallus an' let it drop onta the chain. Well, sometimes the bloke on the winch'd let the door drop afore that wuz prop'ly hooked up an' the chap what wuz doin' it would git caught. We hed a young chap git his shoulder knocked out, dislocated, through that. I ran him inta Workington t' git it seen to.

'I never hed that many accidents happen on board when I wuz skipper, but I can remember a bosun o' mine, Bob Butcher, gittin' clouted b' the trawl one time. This wuz when we were round at Fleetwood. The net come up wi' one o' them gret big baskin' sharks in it, so we got two or three rope becketts round it, an' a wire beckett, an' we hooked the double tackle an' the gilson in an' we hove it up. Well, Bob wuz the one what undid the cod end knot, so he got inta the pound right near the net. I kept tellin' him t' git aft because of all the weight there wuz in the trawl, but he dint take no notice. "I'm all right, skipper," he say. Well, when we'd got the net hoisted right up, the boat give a roll an' the net swung over. The tail wuz stickin' out o' the net an' that caught him an' laid him out. He brooke his arm an' his wrist an' we hetta run him in f' treatment.

'I always used t' tell my blokes t' be careful 'cause there are so many things what can happen on board. I mean, you used t' hear about fellers who used t' git their oilies caught in the winch an' rolled round the drum. Some o' them used t' git really mangled up. Now my brother, he hed a second engineer what lorst a leg goin' down the engine room steps. He wuz on deck, yarnin', when the boat come fast. O' course, they rung down from the wheelhouse t' stop, so this chap made all haste t' git down below an' went an' caught his leg through the ladder's rungs. That wuz broken so badly they hetta cut it orf below the knee. Another thing what happened down in the engine room wuz carbide explosions. You used t' hev a carbide generator at one time f' yuh lights, an' there were one or two trawlers round at Fleetwood what blew up 'cause the drums o' carbide wun't stored prop'ly. Soon as the dynamoes come in, what worked orf yuh engine, that all stopped.

'But apart from that sort o' thing, there were the little things what could happen. I mean, you could slip over on deck an' hurt yuhself. When I first went t' sea, we hed the ow leather sea boots. They were slippery, especially on them iron-deck standard boats what'd bin built durin' the First War. Cor, they were terrible! An' another thing about 'em wuz the noise. When you were turned in, in yuh bunk, you could hear the blokes up on deck a-walkin' about above yuh. They built some of 'em wi' wood above the iron, but that made 'em s' heavy. Oh, they were slushy things − specially when you were shootin' the trawl. They'd roll about all over the place then. But with a good bloke on the winch, he'd slack it out just right. An' once you'd got the bobbins just below the rail, you were all right. They couldn't swing back in an' catch yuh then.

'Course, the weather wuz a thing you hetta watch. The ow Irish Sea could git bad. Round Bailey an' The Kish, you used t' git some bad blows there. You'd git close t' the land if yuh could an' hope f' daylight t' come so that'd ease down. Orf Milford wuz another bad place. That it wuz. I remember gittin' caught there once when I wuz in the *Amalia*. We'd got just past The Smalls, near The Hats an' Barrels, when that come on a gale o' wind. The Hats an' Barrels are rocks. Yeah, there's the Smalls Lighthouse; then there's The Hats an' Barrels; an' then there's Grassholme an' Skokholm an' Skomer, the different islands. Cor, thass terrible there. You git s' much sea a-runnin'. Anyway, like I wuz tellin' yuh, we got caught. There wuz us an' a couple o' big Milfordmen a-dodgin'. I kept lookin' behind. "Oh heck!" I thought t' m'self. I tell yuh, I wuz right glad when that come daylight an' the wind eased, so we could run inta Milford. You git a sou'-westerly round there an' the swell is terrific. Some o' the better boats could do a bit o' broadside in nasty weather, but you daren't do that in the *Amalia* − she wuz such a bad sea ship broadside on she'd ha' likely turned over. All you could do wuz keep her hid to it an' dodge. Some o' the boats used t' dodge inta Milford stern first. I've done that m'self. Oh yeah.

'Where that wuz more open, you could run afore the wind. That wuz like when you were down orf The Longships or The Wolf. You used t' git these big seas, but I never minded. As long as yuh hed a boat with a good stern, you could run afore it. You could hear the ow propeller a-comin' out o' the water sometimes wi' the lift, but that dint matter − you were a-steamin'. Course, you could git these unlucky seas sometimes. You know, freak waves an' that sort o' thing. There's a bad place betwin Scotland an' Ireland where yuh git what they call these overseas. Thass where yuh git a meetin' o' tides. You git that betwin the north o' Ireland an' the Mull o' Kintyre. Once yuh leave the Mull, there's all tides meet an' you git a lot o' turbulence. So if you were goin' north, you'd always go through the sounds if that wuz bad weather. You'd run through past the Mull, goin' down, through Gigha Sound, an' you'd keep all the way inside so yuh passed Tobermory an' Oban. You were all inside. See, you'd got the islands t' shelter yuh. Then you'd come out at Ardnamurchan Point, an' you could either go t' Coll or go betwin Eigg an' Rhum. I always used t' do that, go down through the sounds t' cut out them heavy seas a-runnin' in from the North Atlantic. Then you could run right the way down t' Stornoway or the Butt o' Lewis. Now orf the east coast here you hent got the deep water; you hent got the vast Atlantic swell rollin' in. What yuh hev got is these nasty short seas. Them an' the sands. Yeah, you've gotta watch them.

'Like I said before, you dint hev many things t' help yuh at one time when yuh were trawlin'. I mean, I think that wuz about 1934 afore we hed a radio put in the *Amalia*. I went in her September 1932. She wuz layin' at Milford an' I went round t' join her there. We mainly worked out o' Fleetwood, Milford an' Padstow. We did come hoom one or two summers, but after about 1936, 37, we never come hoom no more. We kept continually round there. We hed an echo-sounder an' all, yuh know. That first one we hed wuz a rum thing. You used t' hetta pour water out o' a waterin' can inta the top of it t' git it t' work! But then they gradually improved 'em so they got better an' better. The real improvement in trawlin', though, wuz after the war when the Decca navigator come in. There wun't no need t' put down dans then t' mark yuh ground. You knew exac'ly where yuh were an' what yuh were doin'. Oh, I liked the Decca. The only trouble wuz that it made fishin' too effective. Yeah, that it did − specially where the foreigners were concerned. When the French, the Belgians an' the Germans come round the west side after the war, they'd foller yuh round. They'd know that you were gittin' a fishin' an' you only wanted t' go over the ground once afore they'd got the job weighed up. They fished Morecambe Bay out as though that'd bin swept by a broom an' they did it just by follerin' the English boats round.

'That used t' cause a lot o' bad feelin', that did, when they come crowdin' in. An' I'll tell yuh another thing what used t' git some of us riled. You used t' git these foreigners fishin' in the Clyde there inside the limit. You daren't go in, but there'd be French, Germans an' all sorts a-fishin' away. Cor blimey, that used t' make yuh mad! They knew it an' all. That they did, 'cause they used t' lower their trousers an' show yuh their bare behinds over the rail. Oh, the French an' Belgians used t' do that a lot. You used t' git that orf Peel, the Isle o' Man, an' all. Well, so yuh did out here in the North Sea. I hev heard my father say that in the real ow smack days one ow Low'stoft skipper used t' carry a shotgun aboard an' he used t' let go with it. You know, he'd hev a pot at these here Froggies an' them if they showed their backsides. I can always remember my father tellin' me that yarn.

'Another thing what used t' happen out here in the North Sea at one time wuz that the Dutchmen used t' run a broom up their forem'st. That put yuh in mind o' the Battle o' Sole Bay, when the Dutch admiral put the broom up his mast t' show that he wuz goin' t' sweep the English orf the sea. So the English admiral run a whip up t' show that he wuz goin' t' whip the Dutch orf. I can remember Dutchmen in the steam trawlers a-wavin' a broom at yuh when they saw yuh, an' yit they were real nice fellers when yuh got t' know 'em. When I wuz in the *Ouse LT572,* I used t' hev a yarn with a Dutchman in a boat called the *Horn Reef.* This'd be when we were fishin' down on the Amrum Bank an' orf Jutland. You used t' be anywhere down from Skillen t' The Elbe, an' if you saw each other, you used t' draw up alongside each other an' hev a yarn if yuh could. That wun't be when you were both towin' along, o' course, but if one of yuh wuz runnin' orf you'd go an' hev a shout.

'With fishin', things are always interestin'. Every trip is an adventure, whether you go down t' the north o' Scotland or anywhere. I mean, no matter where I wuz, I used t' like t' find gulleyways. Where there's gulleyways, there's fish. If yuh could tow through some o' them at the break o' day an', say, haul about eight o' clock in the mornin', or shoot about four o' clock in the afternoon an' haul at half-past seven or eight, they were t' me the best times f' the swimmin' fish. Yeah, they were the best times f' cod, haddicks, whitins, hake an' even dorgfish. Once or twice we've hit dorgfish an' the trawl wuz so loaded with 'em that it nearly burst. You used t' git that in Morecambe Bay sometimes wi' whitins. They wun't very big uns either an' the blokes used t' git fed up 'cause they dint fetch right a lot o' money. "Cor, skipper," they'd say, "let's git orf these! There's s' much guttin'." Course, wi' the French, whitins used t' be a prize fish. Oh, they used t' think a lot o' them.

'Wherever yuh went, you needed t' know what yuh were doin'. Thass why I always listened t' the old men. They could teach yuh a lot. There wuz an ow watchman round at Milford, he musta bin over 80, an' he say t' me one day, "George, if ever you git t' be skipper, hev a go orf Gynfelin Patches Buoy. Run her half an hour on the dead tide north, put yuh dan down an' keep round." Well, I always remembered that an' when I went skipper, I give it a go. Yeah, we went there orf Aberystwyth four or five summers runnin' an' got two or three trips good fishin'. Mixed bags. You hetta watch it though 'cause there wuz a lot o' that honeycomb rock there, coral, an' you could soon split yuh net or smash yuh doors if yuh wun't careful. Thass like orf the Isle o' Man, near The Chickens, you hetta watch it there an' all. The bottom run out gradually there at first an' deepen t' about 22 fathom; then that suddenly drop down t' about 100. We used t' call that The Cliffs. Some o' these skippers what used t' try an' foller yuh in would git it wrong. They'd come an' shoot an' try t' tow in towards the land. They used t' lose all their gear! They used t' hit the edge o' the cliff solid an' that wuz all blue clay. Solid. I used t' like t' go as close t' the edge as I could 'cause, like I say, where yuh git holes an' gulleyways you git fish.

'Workin' the tides wuz another thing. Thass easy wi' Decca 'cause yuh know exac'ly what you're doin' an' where yuh goin'. Afore that come in, you hetta weigh things up a lot more carefully. Yeah, I mean, sometimes you'd drop yuh dan over an' the tide would pull it under − that'd be runnin' that strong. Well, you hetta make allowances f' all that sort o' thing. Thass like if there were wrecks about. Where yuh got wrecks, there wuz always fish, but you hetta be careful. If you were towin' with the tide, you'd leave yuhself enough room t' turn away from the wreck; but if the tide wuz aginst yuh, you could praps tow nearly up to it. If yuh did happen t' come fast, the best thing t' do wuz screw the brakes up on the winch an' wait f' the tide t' turn. With a bit o' luck you'd git orf with hardly a mesh broken.

'Down on the Codlin' Bank, which is orf Dublin, there used t' be a funny ow tide run there. That'd carry yuh along all right! I know one year down there we were gittin' nice dead tides, so we towed right inta the edge o' the bank. The doors used t' come up wi' sort o' white, chalky stuff on 'em. Yeah, that wuz nice fine weather an' we were towin' right in till we stopped. Then we'd swing away on a bit o' cross-tide an' haul. Yeah, I got that taped orf in fine weather an' the ow bags o' cod were comin' up a treat. Course, you hetta study what yuh were doin', but the fishin' wuz there if yuh went after it. Some o' the chaps used t' foller each other round, but I never believed in that. I used t' like t' go on me own. I never wanted t' see boats ahid of me when I wuz out − specially when I wuz after long stuff. No, I wanted t' be alone. That used t' be a joke round there about me bein' on me own. "Where's Greta Garbo?" they used t' say. "Where's ow George?" But, there yuh are, that wuz my way. I always used t' weigh things up an' plan as best as I could. An' yit I wuz no lover o' the sea, yuh know. No. My brothers always used t' say, "He'll never make a fisherman. He'll never be no good at sea."

'I orften laugh about that now. But even though I never really liked the sea, I wanted t' do well. That wuz why I always paid regard t' anyone what hed somethin' t' show yuh. When I wuz on the *Ouse,* that wuz the first time I'd come acrorss the French trawl. There wuz a Frenchman come out with us t' show us how t' work it. Well, I went mate f' one or two trips 'cause the reg'lar mate wun't go while there wuz a Frenchman on board. That dint worry me; he hed suffin t' teach yuh. Thass like when I wuz in the Crown boats − we hed a Frenchman come out with us on one o' them t' show us how t' work the bridle gear. Heck if ow Sam Horn, the mate, dint go on! You know − "I ent hevin' no bloody Froggie tellin' me what t' do!" But thass the way yuh learn. That it is.

'Now I used t' like the French gear. Them bridles give yuh a better sweep.

'Course, I spent over thirty year round on the west side. I think I got t' know the Irish

Sea as well as this room we're sittin' in. That wuz a diff'rent kind o' fishin' in many ways t' the North Sea, but certain things were the same — the east wind f' one thing. I expect you've heard this before, but an ow easterly wind is the worst thing there is f' a fisherman. You dun't git nothin' when thass a-blowin'. Well, thass not quite true 'cause yuh did used t' git the cod in Morecambe Bay when there wuz an easterly. But they're a swimmin' fish. Thass the ground fish what dun't like an easterly. The ow plaice used t' try an' burrow down in the ground an' the white part under their noses used t' be right red an' chafed. We used t' call 'em red-noses, an' you always used t' git 'em after there'd bin an easterly blow. The ground fish dint like a northerly very much either, but that wuz just the job f' fishin' the north end o' Morecambe Bay, along Ross. That seemed as if the fish used t' collect there t' git out o' the way of it. Westerly winds were good f' fishin', generally speakin'. So were southerlies an' sou'-westerlies.

'One thing I did like t' see when we were out on a trip wuz the little ow will-ducks. They're a small bird an' they dive after fish. You used t' git flocks o' them an' that wuz a good sign. Then you'd see the ow gulls a-divin', an' gannets, an' you'd say t' yuhself, "Right, let's hev a look." If you saw birds like that, you knew there wuz somethin' about — an' 99 times out o' a hundred you were right. That wuz what come with experience. You got t' know what t' look for. Like along by Great Ormes Hid: you knew exac'ly where t' go t' git the plaice an' the thornyback rooker. You'd even go inta the Mersey when things were right. Yeah, that yuh would — right in where the big ships used t' run in an' out. Mind yuh, you'd git out o' the way if that come on foggy, in case yuh got run down.

'What I always used t' do when I wuz skipper wuz write things down in a book. I used t' use one o' the diaries what Alec Keay used t' give the skippers in his orffice. Yeah, I'd write things down in that like winds an' tides, an' what the bottom wuz like, an' I'd keep these books from year t' year so I could always refer back if I wuz a bit dubious. "Well, thass the right time o' year, same date, we'll hev a bash." Course, they're like fruit an' vegetables t' me, fish are. You know, they all hev their seasons. So that paid yuh t' study an' weigh things up. An' another thing t' do wuz keep the same crew, if yuh could. A good crew is worth a lot. I mean, if you hed blokes what'd bin with yuh a long while, they got t' know the grounds as well as you did. They could look out f' things when they were on watch; you dint hefta tell 'em.

'We used t' fish all over. I hed my favourite grounds, o' course — I mean, I always liked it orf The Chickens — but I'd give anywhere a go where I thought we might git a trip. If things got a bit scarce summertime, we used t' come up as far as Cardigan Bay an' Caernarvon Bay. I used t' work Caernarvon a lot if there wuz easterly winds. Well, then eventually that all got fished out, so away we used t' go further north, down the North Channel an' orf Barra Head. We even used t' git down t' Cape Wrath, on the north o' Scotland, an' work the Outer Hebrides an' them places. Yeah, we used t' work down there. We'd do a bit o' poachin' sometimes. You know, go inside an' git a tow in.

'I got had up once f' doin' that. Yeah, I got took inta Campbeltown. That wuz when I first went skipper o' the *Amalia*. The skipper o' the *Diana LO31,* one o' Hewett's boats, say t' me, "George, you wanta go down the Clyde. There's good fishin' in there." So I thought I'd give it a try. Well, when I got down there, there wuz Frenchmen an' all sorts a-workin'. Providin' they kept three mile from land, they were all right. But we wun't allowed t' look at it! No, the English boats hetta keep out. Yit these foreigners could go in, run round Ailsa Craig an' come back out agin! Where we were concerned, there wuz a line drawn acrorss from Corsewall Point t' the Mull o' Kintyre an' we wun't allowed t' go inside it. That never seemed right t' me. Anyway, I tried it this once an' I got caught by the Scotch fishery boat. I argued about it in court, but that dint do no good − I got fined £30. I wuz after prime fish in there − buts an' brills an' turbots.

'One or two o' the "Ocean" boats from Yarmouth what used t' go inside there hed electric lights, an' they'd put them out once they were in. We hed acetylene lights in the ow *Amalia* an' we couldn't put them out, so they hetta stay on. We'd only got there that night when I got caught an' the fella what come aboard, he say t' me, "Skipper, why wun't you a freemason?" I knew what he meant all right; if I hed bin, I proba'ly wun't ha' got pulled. Course, several o' the Scotch boats used t' hev the Freemason badge on the bows an' in the funnel, yuh know. Yeah, I can remember half-a-dozen Scotties comin' round t' the north o' Morecambe Bay one year. Well, that wuz along by Ross actually an' that wuz where the English an' Scottish jurisdictionaries met. We were fishin' away there, when up come this fishery boat − same one what'd caught us that once in the Clyde. He say, "Skipper, you're pretty close t' the line, aren't yuh?" I say, "What do yuh mean? What about them?" See, there wuz about two or three o' these Scotties a mile inside the limit. You know, out o' this half-dozen what'd come down. An' they all hed the Freemasons' badge on 'em. "Never mind about them," he say. "Thass you we're tellin'." An' I hetta shift. Course, people were always tellin' me I oughta join the Freemasons, but I say like this an' this wuz always my argument: I like t' be free; I dun't want t' be tied t' no one. I dun't disagree with any o' these societies, but I've always wanted t' be just what I am.

'Fishin' give me that chance really, I spose. Once you're skipper, you're on yuh own. You're the one responsible f' what happen. Sometimes you wished yuh wun't! We got bombed by German planes a couple or three times durin' the war. Cor, that wuz frightenin'! The first time that happened wuz in the *Rosalind*. They were bombin' Liverpool an' Belfast that night an' we were just past Langness, the south-easternmost part o' the Isle o' Man. Now I only hed this happen t' me twice in my whole time fishin' an' this time wuz one of 'em − the soot in the funnel went an' caught alight. Well, that showed up like a torch in the dark an' there were these planes up above us tryin' t' bomb us. What we did wuz draw the fires quick an' hope f' the best. Luckily, that wuz nice an' calm that night an' the planes went away after a bit. That wuz funny how that happened, though, that soot catchin' alight 'cause I only hed it happen to me twice. Some o' the ow chiefs used t' clear the funnel o' soot by hangin' a chain down it. They used t' git half a titler, tie it t' the handrail on the casin' with some wire an' chuck it down the funnel. When the boat rolled, that'd move about an' dislodge the soot. Mind yuh, the crew used t' swear about that 'cause as you lay in yuh bunk that'd be clangin' away every time the boat rolled.

'Another time I got bombed wuz when I wuz in the *Tanager H134*. She wuz an ow fleeter. We were fishin' over The Kish an' we were on our own. What they used t' do when there wuz an air raid on wuz black out all the lights, but leave the buoys 'cause, I mean, they couldn't black all them out. Anyway, this time I'm tellin' yuh about, we were on a nice bit o' fish an' hed got our dan down, when a couple o' German planes spotted us. We couldn't see them, but they could see us, an' they come after us. One o' the bombs come down right close to our stern an' parted the warps, an' because we were hiddin' the tide we couldn't git away from the dan quick enough — an' that hed a light on o' course. Well, there wuz one or two more bangs an' then they cleared orf. We were leakin' like anything, but we got the ow gal in all right an' they made us go on the concrete hard in Fleetwood harbour. They stuck 90 odd rivets in her alt'gether. There wuz that number sprung from the shakin' up we'd hed.

'Thass funny how things used t' turn out sometimes when you were at sea. I wuz thinkin' only the other night about the different things as the year go round. Over on The Kish you used t' git these crawlers certain times o' the year. They're brittle an' yit they're always on the go. They're all legs like a spider an' they feel like shell. You could lose yuh gear wi' them, you'd git so many, an' I never wuz one f' fishin' where you got two or three splits f' one good haul. Then in Morecambe Bay, summertime, you used t' git these blobs. They were bloomin' gret jellyfish an' I've sin the trawl full o' them. Cor, we used t' swear about them! You just couldn't work once yuh hit 'em. Out here in the North Sea, in the spring o' the year, you used t' git them — this is a bit rude — Dutchmen's farts. They're like mudballs, a kind of a shell thing wi' little bristles on what stick in yuh fingers. You could lose a trawl with the weight o' them'.

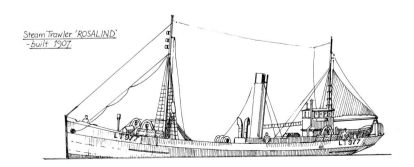

Steam Trawler 'ROSALIND'
– built 1907.

CHAPTER EIGHT

Castle Boats And Crownies
Some Notable Steam Fleets

'Merrily, cheerily, so merry are we.
No mortal on earth like a sailor at sea.
Heave away, haul away, hi-do or down,
Give a sailor his grog and there's nothing go wrong.'
(Traditional − chorus to 'The Seaman's Alphabet')

The steam trawler was long proverbial as a vessel of little comfort and depressing conditions. As we have seen, those of the Crown fleet had a particularly notable reputation. However, the Consolidated Company's Welsh trawlers, known to everyone as the Castle Boats, were much more up-to-date and efficient. Ned Mullender, whom we last met on his frolic ashore at Padstow May Day, was in his time a skipper in both fleets and remembers life aboard them well, particularly one spell fishing out of Swansea:

'I hent bin wi' the Consolidated all that long, when they sent me down t' Swansea. I come in one trip an' ow Mucko Breach, the ship's husband, say t' me, "Mr. Frusher wanta see you, Neddy." I say, "Oh ay. Whass up? What've I done now? Are they goin' t' lay the boat up?" That wuz yuh first thought, yuh know. I f'git what boat it wuz now. That mighta bin the bloody ow *Ostrich LT107* or the *Jonquil GY391.* Anyway, I went in t' see ow George Frusher an' he say t' me, "I want you t' go down t' Swansea." I say, "When?" He say, "T'morrer." Cor blimey, bor! That shook me! I say, "What do yuh want me t' go down there for, sir?" He say, "To teach 'em t' fish f' soles. To show 'em the way t' catch soles."

'Well, that wuz all right, so then I asked him about my crew. He say, "You git them from Low'stoft. You wun't want no engineers, nor yit a cook, but you'll want yuh deck crowd." So out I goes t' try an' git 'em. Well, I asked four or five blokes, but they wun't decide − that wuz too short a notice. See, we were goin' t' leave the follerin' night t' git down there Saturday mornin'. But one chap did say he'd come mate along o' me, so I went back t' George Frusher. I say to him, "I've got me mate an' thass all." I say, "He's the main item. I think the best thing is f' us two t' go down an' sort out the rest when we git there." So thass what we agreed t' do. Well, when I got down t' the orffice the next mornin', George Frusher say t' me, "Your mate's bin in an' chucked his hand in. Go an' see if you can find another one." So out I go. I asked four or five different blokes, but that wuz no go. Then a chap called Billy Mason said he'd come down with me.

'We went down by train an' arrived there at six o' clock in the mornin'. There wuz a car waitin' t' pick us up an' the bloke took us inta a cafe on the docks t' git some breakfast. We each hed a nice big plate o' fried bacon, eggs an' tomatoes, then we went down t' hev a look at the boat afore we met the gaffer. She wuz called the *Henriette GY1326* an' she wuz a nice boat, but the gear wun't no good f' what we were goin' t' do. I say t' Billy, "This ent n' good to us, boy. We've gotta hev a different rig-up t' this." Just as we were comin' ashore, we met a man who wuz all dressed up an' he asked us if we were the men from

Hull steam trawlers pictured alongside in 1933. Nearest is *Amethyst H455* seen in the earlier photograph. Next to her is *Santini,* 96 tons, 56 h.p. built at Selby in 1917, then *Cape Trafalgar H918,* 133 tons, 87 h.p. also built in 1917 for the Hudson Steam Fishing Co. who had a fleet of boats named after famous headlands. The strange structure right aft on some of the boats is a toilet and lavatory compartment for the crew, an addition regarded as a great luxury at the time. It also doubled as storage for spare gear.

Low'stoft an' what did we think o' the gear. So I told him. "Thass not what I want, sir," I say. "Thass far too light." See, the ground-ropes were only just corkscrewed over a piece o' wire. "Well," he say, "I set that gear up an' we do very well with it." He wuz sorta like their main ship's husband round there.

'After I'd spoke t' him, I went in t' see the manager, Major Rollins. He told me that I could hev whatever sort o' gear I wanted, so I went orf t' the store t' tell 'em what I'd need. Well, the first thing I wanted wuz thick ground-ropes – about as thick as yuh forearm – an' I wanted chain t' wrap round them. Then I wanted titlers an' bobbins – an' their bobbins round there were bobbins! Some of 'em were 24 inches in diameter an' there used t' be a big hollow steel one at either end o' the bosom. At the store they said they couldn't git the gear riddy in time f' me t' go t' sea that night, but I said I wun't a-goin' out until I'd got it. In the end they said they'd git me enough t' make up two trawls, as long as I brought 'em some crayfish when I come in. So I promised t' do that, an' the ow ground-ropes an' chain were all aboard by the end o' the afternoon. Well, I wuz pleased an' I promised everybody crayfish.

'When I went aboard the boat that night along o' Billy, the only one there wuz the cook, an' he wuz drunk! All the rest hed gone ashore on the beer. Away come the berthin' master an' he wanted t' know if we were goin' t' catch the tide. I say, "Yis." He say, "Well, you'll hetta git outside the lock." So thass what we did – we got outside the lock an' laid at the jetty just outside Swansea harbour. After a bit, the crew turned up – the chief, the second engineer an' the fireman first of all, then the third hand an' the two deckies. Well, bo'man, I should say; they dint talk about third hands round there. When we got out, I decided that the crew could all turn in while I took the boat t' Lundy. Once we got there, we'd bring up an' git the gear all fixed up, riddy t' fish.

'We hent bin steamin' very long, when the mate come up an' said that the cook wanted me t' go down an' hev a drink wi' the crew. Well, that wuz never a habit o' mine t' drink when I wuz at sea, but I went down f' sociability's sake. "I dun't make a practice o' drinkin' at sea," I say, "but I'll hev a pot o' beer wi' yuh, an' then thass it." So thass what I done an' then they turned in not long arterwards. I hent bin up in the wheelhouse all that long, a-steerin' along, when the chief come up. He wanted me t' hev a drink wi' him! So what I done wuz take the bottle he hed an' just hev a little sip out of it. Then he went down below. When we got t' Lundy, the second engineer wuz on watch, so I rung down an' told him t' stop the engines an' git the crew out.

'We dropped anchor an' got all the lights on – she wuz all electric, see – an' started t' git the gear riddy. Well, bor, that wun't long afore the bo'sman started t' say we wun't doin' things right an' he got a bit argumentative. I say to him, "Never mind what you usually do. While I'm here, you do as I tell yuh. An' if yuh dun't like it, you know what you can do when we git in the harbour." Anyway, we got the trawl rigged up an' away we goes. Well, that wuz a couple or three tows afore we got any quantity o' fish 'cause the bobbins wun't saturated at first an' wun't keep down on the bottom very well. But then we started t' do all right, so I decided t' keep on this bit o' ground. On the third day the fireman come t' me an' reckon he'd hurt his chest the trip b'fore, an' he wanted t' know if I'd take him in. Well, I thought he might be tryin' it on, so I asked the chief about him. "Dun't you pay no regard t' him!" he say. "When he's dead, chuck him on the ice. We can take him in then when the trip is up."

'Well, that wuz good enough. That wuz his way o' tellin' me what this fireman wuz like. So we carried on fishin'. On the fifth day up come the fireman agin. "You've gotta take me in, skipper," he say, "or I'm goin' t' report it when we git in." Well, I got a bit worried then 'cause he mighta bin hurt, an' if he got worse that wuz my responsibility. So we run back t' Swansea an' got in on the night tide. The berthin' master told him t' report t' the hospital an' I even give him some money f' his fare. A bob, I think it wuz. O' course, nexta mornin' we landed an' we made about £130, which wun't bad f' the time we'd hed fishin'. An' there wuz plenty o' crayfish f' the blokes on the store! I took 'em round there an' dished 'em out. Well, bor, you'd ha' thought I wuz givin' 'em a fortune! Now the fireman, he wuz tryin' it on. Yeah, that he wuz, an' he got the sack. He went an' complained t' the union about it an' the union people set tight on the gaffer first of all — on Major Rollins. He sent 'em t' see me an' I told 'em what he'd bin up to. An' the funny thing wuz that I wuz a union man an' all. Yeah, I belonged t' the same union in fact! Seamen's branch o' the Transport an' General Workers.

'After that wuz all cleared up, out we go agin an' carried on fishin'. We dint stick round Lundy all the while. No, we used t' try the various places. We used t' go down t' The Wolf an' work orf there. Then we'd git t' The Seven Stones an' give that a try. We dint used t' go acrorss t' the Irish side, though. No, we worked all this side. We used t' run inta Morecambe Bay an' fish there. Yeah, that wun't bad in there. You could git some nice trips. Soles were the main item you were after, but you got other stuff an' all. I mean, one trip we come in an' we had soles, plaice, dabs, lemons, witches, whitins an' a few hake. I suggested t' Major Rollins that we fish a trip down an' land in Fleetwood, then fish a trip back an' land in Swansea. Like that, you'd git a day or a day an' a half extra fishin', 'cause that used t' take yuh about 36 hours t' steam either way. He say t' me, "Thass a good idea, skipper, but we want the sellin' commission ourselves." See, they were sellin' their own fish an' gittin' the commission, so they dint want you t' land anywhere else apart from Swansea.

'Gittin' inta May, things began t' git slack, so I thought I'd better find a new fishin' ground. I went up t' one o' the ol' grounds I knew about, pretty nigh up t' The Seven Stones, an' we fished a decent trip. We made about £200 f' the eight days. That wuz mixed stuff — rooker, a few hake, a few soles, whitins an' dabs, an' a few plaice. The next trip wuz a flop. We couldn't find a bit o' fish nowhere, so we stopped out about ten days. When we come in, I think we only made about £90, so Major Rollins decided t' lay the boat up f' a bit. This wuz the *Henriette*. She wuz a three furnace job, so she used t' burn a fair bit o' coal. "Yeah," he say, "I think we'll lay her up for a bit, skipper. You hent paid this trip an' things are gittin' a bit slack. And," he say, "We lorst enough money on her afore you come." So that wuz what happened — they laid her up. I decided t' come hoom, but he say t' me afore I went, Major Rollins did, "If you decide t' come back agin, we'll give yuh a berth in one o' the hake boats an' teach yuh t' fish f' hake."

'Hoom I come. Well, there wuz nothin' doin' here, in Low'stoft, so after about three weeks o' tryin' t' git fixed up I say t' the wife, "I'm goin' back t' Swansea." An' thass what I did. I fixed m' self up wi' digs an' they give me a bo'sman's berth in the *Powis Castle SA68*. She wuz a big boat, about 130 foot on the keel, but I wun't in her long 'cause me an' the skipper couldn't git on. That wun't no one's fault really. He just dint want me 'cause I'd bin put there in place o' one o' his reg'lar crew. Anyway, after a bit, they put me in the *Harlech Castle SA42* along o' Jimmy Green. He come from Corton originally, but he'd bin down there quite a long time. I went along o' him f' about four trips, after hake. We were workin' the French gear, with about 60 foot bridles, an' that wuz very light compared t' what I wuz used to. The bridles used t' run through the doors, yuh know, through a hole, an' the firm wuz payin' a royalty t' the French Gover'ment f' bein' allowed t' use it. Well, after a while, someone invented the pennant; so instead o' yuh bridle runnin' through a hole in the door, that wuz joined t' the warp (when you were haulin') by this pennant, an' that meant everything could run onta the winch without touchin' the door. An' that finished payin' the royalties 'cause that wuz a diff'rent method.

'After I'd done them four trips in the *Harlech,* Major Rollins put me in the *Pointz Castle SA5* as skipper. The first trip we went down t' The Labrador Bank an' the Adelaide Bank, sou'-west o' Ireland, down on the coral, an' we fished a decent trip. We made about £220 f' the 12 days, so that wuz a payin' trip. When we went out agin, I say t' George Wright, my mate, "I can remember gittin' hake orf Pendeen when I wuz along o' my father in the drifter-trawlers afore the First War. I think we'll hev a try down there." This'd be the end o' July, beginnin' o' August time, so we went down outside The Cape Cornwall Bank. We were about 22 mile west-nor'-west from Pendeen, in betwin two roughs o' big ow heavy stones. I hed two or three casts o' the lead t' see what the soil wuz like, then we dropped a dan down an' shot away. We found the hake all right. In 48 hours we'd got a trip!

'Now I wuz supposed t' send in what they called an armadale every night t' say what fish I'd got. That wuz a code name an' yuh sparky what wuz on board, he used t' pass all the information inta the office. He wuz a proper sparky, employed by Marconi, an' they used t' pay him about four quid a week. On top o' that, he used t' git £1 in the gross hundred on the trip's earnins, which wuz like a bonus for him, so he used t' do all right. He used t' send the information by morse code, but when we were on all these hake I hed him reduce it right down t' only a little 'cause I dint want anyone else a-comin' after 'em. See, there wuz always other boats a-wiggin' in on your signals an' a lot of 'em knew the code 'cause they'd bin in Swansea ships. An' they'd hev their direction-finders goin' as well t' try an' pin yuh down. Well, I dint want them droppin' onta me when I wuz on a good livin', so I lied about what we were gittin'.

'I give it away in the end because I heard a chap what'd bin along o' my father afore the First War talkin' over the radio. Tiddly Mortimer his name wuz, an' he wuz a skipper in a steam trawler that wuz fishin' down that way. I told him we were on a good livin' an' he say t' me, "Cor, you're some loud! You must be fairly close to us." I say, "Yeah. Git yuh direction-finder riddy." So thass what he done; he DF'd us. About an hour afterwards he come up alongside. "What're yuh gittin'?" he say. So I told him – "Hake," I say. "Good God!" he say. "How long ha' yuh bin on them?" I laughed. "Oh, about 48 hours," I say. "Hurry up an' git yuh gear down, 'cause someone's undoubtedly heard us." Well, anyway, we both hed about four or five short hauls each an' then the crowd come. Away they come from all directions! Frenchmen, Belgians, the lot! But I dint care 'cause I'd got a damn good trip aboard an' Tiddly hed got a full day's work in. You oughta sin 'em, though — they soon scored that bit o' ground up. When I hed the sparky call the orffice the next time (I wun't allowed t' send morse, yuh see; that wuz his job), I got him t' bump the catch right up an' they called us in. When we landed, we made over £700, which wuz the biggest trip in there f' a long time.

'Course, the French gear we used wuz a lot lighter than the ordinary otter trawl. The heaviest bit o' the work wuz when you pushed yuh bobbins over the side when you were shootin', though, o' course, you dint use bobbins f' hake. You just hed a corkscrew ground-rope. There's a lot o' rough ground round on that west side, yuh know. I mean, if you git anywhere south or south-west from The Tuskar, you find yuhself on the coral an' you just can't work there. That ruin yuh blinkin' net. Hake are a daytime fish an' they like a roughish soil. The other thing about 'em is that they're a swimmin' fish, so you want yuh gear fairly well up orf the bottom. We used t' hev about 150 bottles along the hidline in what we called sausage nets. They were long, narrow nets all joined t'gether an' you used t' put yuh bottles in them an' then tie 'em along the hidline. After a while, the bottles used t' git water in 'em. Yeah, they'd fill up wi' the pressure, so you used t' break 'em on the rail o' the boat then an' hull 'em over the side.

'You used t' tow f' about six hours when you were after hake, so you'd git two hauls in most o' the year, but three hauls durin' the summer. While I wuz round there at Swansea, that wuz all soles an' hake. Specially hake. Thass the lovable fish round there. Well, so it is at Fleetwood an' all. Course, the thing about a hake is that you wanta doctor him up a bit when yuh cook him. Baked in the oven, in gravy, thass the way. An' another thing about hake — they're a fish whass bin punished. I mean, so many people ha' bin after 'em an' they've crippled the fishin'. Specially the Spaniards, with their pair-fishin'. You know, two boats pullin' one blinkin' gret trawl.

'Another thing you used t' git round there wuz ling. I remember one trip in the *Powis Castle,* we got a bag o' ling you coulda walked on! There wun't a lot o' demand for 'em as fish, but they were full o' rows — an' rows were stockie bait! You used t' git conger as well. They were all right 'cause you could send them t' France. Pollack wun't much cop, though. No. Nor were them ow blue skate an' the thornyback rooker. The blond rooker were all right, though. If you could hit a patch o' them, you'd make a bob or two. Nowadays, o' course, everything make money. I mean, there's a vast diff'rence betwin 1930 an' now.

'Sometimes we used t' go down onta the Labrador Bank an' then work up round the west coast o' Ireland. At night time we used t' run inta the Bay o' Killeany, in Galway Bay, an' lay there. Thass when we used t' do our mendin' an' replace the bottles what'd filled up wi' water. I hev bin out as far as the Porcupine Bank. Thass right out in the Atlantic. If the weather come on too bad t' work, you used t' either heave to, if yuh could, or head the boat up t' wind'ard an' keep dodgin' her. The reason I come hoom from Swansea wuz because I got hurt out on the Porcupine. I wuz helpin' the crew stow up the trawl 'cause the weather hed come on nasty. I wuz standin' near the wheelhouse an' the lashin' I hed hold of broke. The sea washed me overboard an' then that brought me back agin. I got washed up aginst the little boat an' that give me a real ow bashin'. I wuz bruised all over an' they hetta bring me in. The doctor put me on the club, so I come back hoom t' Low'stoft.

'When I wuz better, I went in the ow Crown boats agin. Yeah, I wuz in several o' them alt'gether. I hed the *Aberdeen LT123* f' quite a while. She wun't a bad little ow boat 'cause she wuz very light on coal. She bunkered enough t' last 12 or 14 days without bein' full up, so I used t' be able t' run down t' Skillen Corner an' Borkum an' all them places. You'd git mixed hauls there – rough stuff, haddicks an' plaice – but a lot of it wuz on the small side, so that dint pay yuh t' keep there all the time. I used t' go an' hev a go in the Botney Gut quite orften, an' in the South Rough as well. I never bothered much wi' the Horspital Ground, though; the fish there always seemed t' hev scabs on 'em. I dun't know if thass how the ground got its name, but you did notice these little sores on the fish. Mind yuh, they were all right t' eat. Yeah, I mean, we used t' hev 'em f' breakfast when we did go fishin' there.

'I hev bin down as far as the Monkey Bank. Thass well down, that is. Orf Denmark. But them sort o' trips were unusual. Yeah, I mean, a lot o' the time you'd be out here. I remember ow George Frusher sayin' t' me one November that he wanted me an' another skipper t' hev a go out here. This wuz when I wuz in the *Aberdeen*. Well, we run up t' The Gabbards an' worked there wi' the ordinary on-the-door gear. Course, you always used t' work starboard side if yuh could; the bind o' the propeller used t' draw the net in towards the boat on the port. Mind yuh, yuh port gear wuz always there t' work, if yuh wanted it, 'cause there were times when it wuz handy t' hev it. This time we were up The Gabbards, we dropped onta some soles, but we wun't gittin' anything like the quantity we shoulda done. Where we were gittin' about a weighin', a weighin' an' a half, on our dark hauls, the other boat wuz gittin' three an' a half t' four. Well, I couldn't make this out at first, then I decided t' check the titlers 'cause we'd hed new ones put aboard afore we come out. When I hed a look at 'em, they were far too long. That meant they were travellin' in under the bobbins, see. Yeah, they wun't a-scourin' the ground up like they shoulda done. So we cut 'em down t' the proper size an' we fished lovely after that.

'That wun't too bad workin' f' the ow Consolidated. The main trouble wuz the way they used t' lay the boats up February, March an' April time. Things got slack round about then an' they used t' lay about half the boats up. They used t' take 'em an' put 'em in what they called The Crick – thass near Morton's factory. In order t' keep a-workin', you hetta earn an average of £13-10-0d a day over the whole trip, an' that wuz pretty difficult that time o' the year. But if you dint make it, they'd lay the boat up an' stand yuh orf. Thass one reason why I left 'em in the end. I wuz skipper o' the *Leeds LT131* in 1936, but I left t' go mate in the *Gula LT179*. She wuz one o' the Ice Comp'ny diesel boats, a motor trawler, an' I wuz goin' t' be just about as well orf goin' mate in her as I hed bin skipper in the *Leeds*. See, I got the same poundage near enough an' the expenses wun't s' much, so that helped t' make up f' the smaller share mates got.

'Skippers an' mates went b' the share on trawlin'. A skipper got £3 a week sent hoom an' £10 on every clear £100. A mate got 50 bob a week an' £7-10-0d on the clear £100. Both them weekly payments used t' be taken orf yuh share earnins at the end o' the voyage. All the rest o' the crew were on a straight weekly wage, plus so much in the clear hundred poundage. The chief used t' do best on that – he got 24 shillins, an' the rest were scaled down accordin' t' their position on the boat. The wages were the same: the chief got about 50 bob, the third hand £2, the deckhands about 35 bob an' the cook £1.

The trimmer an' the stoker were somewhere betwin the cook an' the deckies. All the crew got the poundage an' they all got stockie bait as well. In the Consolidated the stockie bait wuz made up o' small jinnies, gurnards an' stuff like that. That wuz doled out the same as the poundage, accordin' t' your position on the boat. The crew wuz allowed £10 f' stockie. If the jinnies an' gurnards come t' more'n that when the trip wuz sold, the extra money went inta the trip. There wuz nine in the crew, so the stockie wuz worth hevin'. Oh, definitely. All the Consolidated boats paid one shillin' in every pound o' stockie bait towards the Low'stoft Hospital. So if yuh stockie made £9, you'd pay nine bob t' the hospital. They hed a ladder in the comp'ny orffice an' each boat's name wuz in a little flag, an' that used t' show yuh your position on the ladder − you know, how much you'd paid inta the hospital. We were top boat one year, when I wuz skipper o' the *Aberdeen,* an' I think the comp'ny that year paid about £400 alt'gether from all the boats.

'Course, the ow Crown Boats hed a name f' livin' conditions. Some o' the crew used t' live down the cabin; the rest were up for'ad in the foc'sle. Yeah, the skipper, the mate, the chief, the second engineer an' the cook slept aft; the rest slept for'ad. You used t' sleep in yuh workin' clo's all the time you were at sea an' you used t' hev a tub f' a toilet. That wuz an ow barrel an' some o' the blokes used t' put a plait o' rope round it t' make it more comfortable t' sit on. That stood anywhere on the lee side − any lee place you could git! Course, the latter part o' the time they began t' put toilet places on the boats. Not flush ones like they hev now, but a little hut thing aft side o' the galley with a seat inside an' a tub underneath − or half an ow carbide tin. No, there wun't a lot o' comfort in them ow boats, I can tell yuh! I mean, the skipper's berth wuz down in the cabin along o' the rest, though them big boats I wuz in round at Swansea hed a separate place f' the skipper under the wheelhouse.

'Mind yuh, you dint live too bad, providin' you hed a good cook. Someone what could make things out o' what yuh hed on board. I know you used t' eat a lot o' fish on board, a fish breakfast every mornin' an' proba'ly a fish tea as well, but beside the fry-ups you'd hev yuh cooked dinner in the middle o' the day. Course, you dint waste any food; you hetta keep t' what you hed on board an' make it last the trip. That wuz no good runnin' short afore yuh come in. Skippers an' mates hetta pay f' their own food, yuh know. They still do. Yeah, the food bill f' the trip wuz divided b' nine, the number in the crew, an' that sum wuz deducted from yuh share. That never really seemed right t' me, but that wuz the way it worked. I remember four or five o' us skippers once tried t' git tax relief on our food bills. This wuz after the war an' we were payin' a fair bit o' tax on our earnins, so we thought we'd try t' git a little back. That dint work, though. No, you never git anything back from that lot, do yuh?'

Mid-water steam trawler c. 1900
70 tons net.

Boston Phantom FD252 covered in ice on an Arctic voyage. This is a fairly mild example of a constant and dangerous menace encountered by the distant water trawlers in high latitudes. Ice forming on rigging and top hamper faster than it could be chipped away might quickly make a rolling vessel unstable.

CHAPTER NINE

Distant Waters

Far North in Search of Fish

'When clouds do slowly onward crawl,
Shoot your nets, your lines, your trawl.
But when they gather thick and fast,
Keep a sharp look out for ship and mast.'
(Traditional — Fishermen's weather rhyme)

All fishing is a dangerous business and trawling has the highest accident and mortality record of all. As the steam trawler fleets pushed further northwards into Arctic waters, so the risks multiplied. Beyond the normal hazards of the job were added the freezing temperatures, black frost, the long periods of winter darkness, the long steam to and from the grounds and the relentless round-the-clock routine of shooting, gutting and hauling to make the trip worthwhile.

The steam trawlers developed for this distant water work, and their immediate diesel successors, were among the most seaworthy vessels ever built. They are never celebrated in those coffee table books of the sea, like the 'romantic' square riggers and the men o' war, but they deserve to be. Sadly, not one steam trawler of the period is preserved for posterity.

These vessels were all conventional side-hauled trawlers — 'side-winders' as they were colloquially referred to. Either as coal burners, oil-fired steamers or, latterly, diesel-engined they were the workhorses of the British fishing fleets throughout the first half of the 20th century. During the 1960's they began to be replaced by diesel-electric stern trawlers which make deck working easier for the crew, but are not to be compared with their predecessors for either grace of line or sea keeping qualities.

Until the majority of the British distant water fleet was laid up after the Icelandic grounds were closed in 1976, the great distant water ports were Hull and Grimsby, followed by Fleetwood. Several Lowestoft fishermen shipped out of these ports to join the search for fish a thousand miles from home in some of the most inhospitable waters in the world. Harry Colby (born 1902) talks about his experiences on distant water trawlers before, during and after the Second World War.

'I think the first boat I went up t' Iceland on wuz the *St. Just H320.* We'd bin workin' the Kildas an' up orf Eagle Island, when the owners decided t' send her t' Iceland. We used t' come out o' Fleetwood an' coal at Stornoway, 'cause you dint carry enough coal t' git all the way up there. You used t' fill yuh fish-room up an' you hed a tunnel through t' the engine-room where you hetta trim the coal through. You used t' git £2 a trip, the two deckies, t' trim that through an' clean the fish-room up after you'd emptied all the coal out. Then, when we got up there, we used t' coal at the Westerman Isles an' you'd do the same agin afore yuh come back hoom. The boat wuz about 140-145 foot long an' she wuz a good sea ship. So wuz the *Wyre Nab FD190* what I went on along o' Runner Harper. Yeah, we used t' run up t' Iceland an' Jan Mayen Island on her. So we did the Faeroes. That wuz all the same sort o' fishin' — cod an' ollabuts.

'When we worked Iceland, the length o' trip wuz about 18 days. That used t' take yuh five days t' git there an' five days back, so you hed seven or eight f' fishin'. When I first went in the *St. Just,* along o' Loopy Sutton, we'd bin workin' about 18 hours all in one go, when I went up for'ad an' the mate say t' me, "Here yuh are, Pud. (They used t' call everyone what comes from Low'stoft Pud) Take this up t' the wheelhouse an' give it t' the ow man. He'll know what it mean." An' he give me a domino wi' no spots on. I took it up an' handed it over, but the ow man dint say nothin' so I come back down agin. When that come t' haulin' time, the cod end come in an' we kept lookin' up t' the wheelhouse t' see what the ow man wuz up to. Soon as we'd dropped the fish out, the mate say, "All right, pull it aboard. Stow yuh trawl, git the fish orf the deck an' turn in." See, that blank domino wuz t' tell the ow man the crew'd hed enough, an' he knew it. He dint argue about it or nothin' an' we turned in. Well, that wuz only right. I mean, when you'd bin on deck 18 or 20 hours you needed some rest.

'In 1936 I went t' Bremerhaven in Germany an' fetched the *Northern Gem GY204* hoom. She wuz one o' Lawford's boats, a big triple steam job with about a thousand horse-power engine. I went up t' Spitzbergen in her. Yeah, we went all that way up there an' all the way back on the same bunkerage. We used t' run the coal in the main bunker up t' the stoke-hole in little tiny trailers. They were on rails like the little trucks they hev in a coal mine an' they used t' tip up t' empty the coal out. You used t' work more or less the same sort o' trawl you did anywhere else. I spose our hidline wuz about 85 t' 90 foot. The only diff'rence wuz that you hed a 12 foot lengthenin' piece in betwin the belly an' batins an' the cod end. That wuz t' give yuh extra room f' when yuh got a big bag o' fish. The net dint hev no flapper on either an' that wuz because o' this lengthenin' piece.

'We dint hev no quarter-ropes on the trawl neither. We used what wuz called smash-and-grab gear. The trawl wuz iron-bound from door t' door, which meant you hed iron bobbins right the way acrorss the ground-rope. When you hauled, after yuh doors were hooked up in the galluses, you hed a strop on the top arm o' yuh butterfly what you clipped onta the bridle so the bobbins could be winched aboard. Not hevin' t' mess about wi' quarter-ropes meant you could shoot an' haul quicker. We hed 15 fathom bridles an' the length o' warp would all depend where you were workin'. I should think the maximum wuz about 220 or 240 fathom when you were in the deep water. What you used t' do when yuh got a big bag o' fish wuz make several separate bags an' drop 'em out a bit at a time. You'd git a becket round yuh cod end, heave up, drop the fish on the deck an' tie the cod end up agin. Once that wuz back in the water, the ow man used t' go astern t' shake down the fish what'd worked their way up the net, then you'd git another bag out. You used t' go on like that till yuh trawl wuz empty. The ow man used t' say t' mate when the net come up on top o' the water, in towards the boat, "What do yuh reckon there is?" An' the mate'd say, "Oh, about eight or nine bags, I reckon, skipper."

'There wuz one trip we hed up there t' Spitzbergen an' we couldn't sell the fish when we got hoom. The cod all smelt an' tasted o' weed an' no one wun't buy 'em. We dint go agin after that. If you wanted to, you could coal at Spitzbergen, but we dint bother 'cause we carried enough on board. An' another thing, that coal up there wuz a bloody load o' rubbish. They used t' mine it theirselves, but that wuz such bad steamin' coal. That wuz all dust. I mean, yuh ordinary furnace bars used t' be fairly wide apart, but if you used that Spitzbergen coal they'd need t' be much closer t'gether so the stuff dint run through. Course, you never carried a deck cargo o' coal on the distant water boats – not in my time, any road. That wuz too dangerous. Yeah, you'd hev all yuh coal down below. You'd hev yuh bunkers right full an' you'd fill up yuh fish-room as well.

'The coal you used t' git at Iceland wuz all right, but o' course that'd come from England anyway. Durin' the war you used t' see "Heil Hitler" painted up in white all over the place in Iceland – Reykjavik, the Westerman Isles, everywhere. Blast, yeah, the Germans never sunk any o' Iceland's ships! They used t' run t' Fleetwood durin' the war wi' fish an' take motor cars, trawl warps an' coal back with 'em. They used t' git the coal cheap as nothin' an' they'd take a deck cargo back with 'em. Then they used t' charge yuh over the odds for it when you put inta harbour. They knew you'd gotta hev it t' git back hoom agin, yuh see.

'We used t' run down there a lot durin' the last war. I went down in the ow *Warbler LO251*. There used t' be about ten or a dozen of yuh in a convoy, with a couple o' large armed trawlers in charge. When the weather wuz bad they could fish, but the rest of us couldn't 'cause we wun't big enough. We used t' fish under orders from the two biguns an' when you'd hed your time, you hetta go hoom – fish or no fish. We hed a little ow six-pounder gun on the foredeck, a machine gun either side o' the wheelhouse, an' a couple o' rockets there as well. When an aeroplane come over, you used t' fire them, an' they used t' go up in the air an' a parachute thing used t' open out with a wire a-trailin' from it. Afore we went out on the boat, we hed two or three days trainin' at the gunnery school in Fleetwood, but I wun't much the wiser for it. (The rocket line was intended as a deterrent to close approach, rather like a barrage balloon).

'You were on yuh own when you were fishin', but you used t' hetta rendezvous back at a certain time. I remember when I wuz bosun along o' Bill Dreever in the *Drusilla A133*, we were workin' the south-east bank o' Faeroe one trip an' you used t' leave guttin' yuh fish till daylight. You darsn't put yuh lights on, yuh see. An' this here particular mornin' we hed a whole deckload o' fish after trawlin' all night, an' we were just steamin' up t' our dan t' lay an' gut, when one o' them blinkin' Focke-Wulfs hed a go at us. He missed the first couple o' times, but he got us on the third run. Cor, dint he come down! Our funnel wuz like a pepper-pot from the bullets an' he dropped some bombs one side of us. After they'd exploded, you'd ha' thought the ship hed bin in dry dock, with a chippin' hammer workin' all along her side. That wuz all like silver plate, right the way along. We hed an oerlikon gun on the casin', but no one dint make a move t' use it. I wuz that frightened I wuz shakin' like a leaf. So were moost o' the crew. We were all up for'ad, a-hidin'. In the end, the mate managed t' git up t' the bridge an' send a message t' Thorshavn. The radio wuz sealed acrorss the terminals an' you wun't supposed t' use it unless you were in real trouble. He got shot twice through the neck gittin' up there, but he survived it an' got a message out. The *Preston North End GY82* an' the *Wolves GY104* come t' see what the matter wuz – big armed trawlers from the Navy.

'I wuz out o' Grimsby an' all durin' the war, workin' Iceland. We used t' go through The Pentlands an' coal at Scrabster, an' thass where we used t' git our orders about where t' go an' where t' fish. We hetta report back there when we come hoom as well. Yeah, there wuz a drifter there called the *Recruit BCK212* an' she wuz full o' naval men, who used t' give you yuh orders. You got inta trouble if you dint stick to 'em. Oh ah, yis, you wun't allowed t' go just where yuh liked. See, there wuz always that danger from enemy aircraft or submarines. We used t' see the ow submarines sometimes, 'cause you used t' break from the convoy when you reached the fishin' grounds. On the return trip, when you were comin' down the North Sea, you used t' hetta stop at these big buoys wi' lights on 'em an' see if there wuz any airmen in 'em. They used t' hev ladders up the side, so if any airmen come down in the water near 'em they could climb up an' wait t' be picked up.

'One boat I wuz in, the *Lord Minto FD51*, got bumped orf by a submarine the trip after I'd left her t' join another ship. They were fishin' out at The Kildas, when this here German sub got her. I think the Germans sunk about five trawlers that night, but the *Lord Minto's* crew were all right 'cause an American destroyer come along an' picked 'em up. There were no end o' boats fishin' out o' Fleetwood durin' the war. They dint hardly know there wuz a war on; that wuz business as usual. The first part o' the war, afore I went deep water, I wuz in the *Regnault H156*, a-workin' Morecambe Bay an' all the other places. We used t' hetta chop the hids orf the fish after we'd hauled so we could carry more on board. You used t' do that when yuh went t' Iceland. These here Navy blokes would hev a look an' see what yuh fishin' capacity wuz, then they'd tell yuh where t' go. Praps they'd say, "Right. You're gotta go t' Iceland." Sometimes you used t' save all the fish hids an' keep 'em in a separate pound an', o' course, you'd keep the livers an' fetch all them hoom agin. They dint allow us t' go t' Norway durin' the war, but when I went through the fjords afterwards I saw the *Tirpitz*. Yeah, she wuz there in Tromso Fjord an' they were cuttin' her up. She wuz a big attraction, she wuz, an' they were runnin' coachloads o' people out t' see her. We went right alongside. Cor, the size of her! She hed four propellers on her an' each one wuz as big as this room.

'When you were on that distant water fishin', the Germans wun't the only thing you hed t' worry about. You hetta be very careful o' the convoys what were runnin' up t' Murmansk. You never knew when you were goin' t' git run down durin' the night. No, that you dint. See, you wun't allowed t' hev no lights or nothin'. You wun't even allowed t' hev a fag on the bridge! I got run down once orf Ireland. I wuz in the *Auk H755* then an' we were runnin' in convoy from Rathlin Island. We were steamin' hoom from Rathlin t' The Maidens, comin' towards Fleetwood, when this big boat hit us. She wuz loaded wi' tanks f' Russia an' wuz just goin' t' join her convoy at Belfast. I wuz turned in when she hit us. There wuz this big bump an' I thought, "Right. Out you git!" I managed t' squeeze through near the mizzenm'st an' the after-gallus an' I jumped onta the anchor o' the big boat. That wuz hangin' down a bit an' I jumped onta one o' the flukes just as she wuz goin' astern. Two officers got me aboard, but I dint stay there long 'cause the *Auk* stayed afloat. She hed this box fish-room, which you could walk right round, an' that acted as a water-tight compartment so she dint sink. When I got back on board, we stuck a trawl net inta the hole where she'd bin hit an' got the ow gal t' Fleetwood. There wuz a fair bit o' water in her up for'ad, but we got her hoom all right.

'Course, I can't swim, yuh know. No, I never could swim. I dint believe in it t' tell yuh the truth. I mean, if you did go overboard out there on that distant water, what chance'd yuh got? The temperature o' the water wuz enough t' kill yuh. That'd stop yuh heart right away. I mean, look at all the blokes what were lorst orf the convoys durin' the war — all good swimmers. I went t' swimmin' lessons when I wuz at Morton Road School in Pakefield, when I wuz a boy, but I never learnt. No, I dint, an' thass a fact. But that never did worry me. Like I say, if you went overboard on that distant-water fishin', you were good as dead anyway.

'After the war I carried on wi' the deep-water fishin'. I wuz with the Boston Comp'ny for a time, workin' along o' Arthur Lewis. He wuz skipper then, but he worked his way up t' bein' in charge round at Fleetwood in the end. After I'd bin wi' Boston a few years, I went back inta Captain Lawford's boats. He wuz one o' the best owners I sailed for. He hed a lovely fleet o' ships an' that *Red Hackle LO109* I wuz in, well, she wuz about the finest trawler ever built, I reckon. We went an' got her new in 1951 from John Lewis's yard in Aberdeen. She wuz an' oil-burnin' boat an' she wuz absolutely beautiful. In the ow coal-burners you used t' hetta climb three bunks high sometimes t' git inta yuh berth, but this one hed all little cabins with just two or three berths in each. An' she hed a proper mess-deck where yuh ate yuh meals. Vera Lynn wuz there when we left port an' she presented us with a couple o' photographs of herself — one f' the mess-deck an' one f' the skipper's berth. When we landed in at Fleetwood after our maiden trip, she sent an invitation down t' the ship for us all t' go an' see her in her show at Blackpool, which we did. That mornin' when we sailed from Aberdeen, we even hed The Black Watch t' play us away! I shall never f'git that.

'We used t' run up t' Greenland on the *Red Hackle* an' we used t' work the White Sea as well. She wuz a four furnace job an' she used t' bunker nearly 400 ton o' fuel oil when she wuz full up. An' on top o' that you used t' carry about 70 ton o' fresh water f' the boilers. You never put no salt water in her, not like yuh did in the ow coal-burners. She hed four big fish-rooms on her an' there were a couple o' machines on the fore-deck what washed the fish afore they went down below. They used t' hold about 15 kit each, I should think, an' after the fish'd bin through them they used t' drop down through a gap in the false stage onta the fish-room floor. See, you hetta hev six foot o' air space betwin yuh hatch cover an' where yuh finished icin' orf yuh fish, so you hed this false stage which wuz made o' movable boards. As soon as the fish-room wuz full, you used t' batten down the hatch cover an' put a waterproof cover over it. Some o' the ow coal-burners I wuz in used t' hev a cowhide go round there t' make it watertight an' you used t' wire it on t' make sure it held.

'There wuz 370 aluminium pound boards alt'gether in the *Red Hackle,* an' they all hetta be scrubbed clean wi' wire brushes. There wuz three fish-rooms f' bulk fish, where everything wuz lumped in t'gether, an' one wuz kept f' shelf-fish. They were the big haddicks an' cod you got in the last two or three days afore yuh went hoom. All yuh fish wuz laid on trays, but apart from them haddicks an' cod what you shelfed, you dint separate anything out. Not even yuh ollabuts, an' they used t' fetch a good price. So did the livers, but you couldn't git the oil out o' them like you could the cod livers, so you used t' put 'em in trays an' fetch 'em hoom wi' yuh. They used t' press 'em t' git the oil out, where cod livers are boiled. We hed a machine on board t' do that, but we never put the ollabut livers in it 'cause they go hard as a rock if yuh boil 'em.

'On nearly all the near-water an' mid-water trawlers the mate is in charge o' icin' the fish. But not on the distant-water boats. What he'd do wuz pick out three blokes as you were runnin' down an' ask them if they were willin' t' take charge o' the fish-rooms. Course, he'd only ask chaps he knew he could rely on. I mean, you could soon ruin a trip if yuh dint know what you were doin'. When yuh landed, if you'd done all right, them blokes what'd bin icin' the fish away used t' git praps ten or 15 quid betwin 'em f' that extra responsibility. All the time you were guttin' up, they used t' be down below icin' away an' they dint come up till all the fish were orf the deck. You knew how much fish you hed down below 'cause that wuz marked up in the wheelhouse every mornin'. The *Red Hackle* wuz a refrigerated boat. She hed a motor up under the foredeck t' work all that; an' when yuh landed, you used t' leave the catch t' defrost for about 24 hours afore that wuz unloaded. Oh, she wuz a beautiful ship. I mean, them washin' machines, they used t' save no end o' work. There wun't no messin' about, washin' the fish by hand an' then handin' 'em down in baskets, like there wuz in the ow coal-burners.

'When yuh went t' Greenland, you used t' work orf Cape Farewell, an' if the weather come on bad you used t' go up the Davis Straits t' look f' shelter. That dint do t' tow too much in the dark up there 'cause there wuz a lot o' icebergs about an' you hetta be careful o' them. We only hed galluses the starboard side. Yeah, she wuz only a one-sider; all the rest wuz sleepin' accommodation. See, we carried a double crew in her, 22 men. After the skipper an' mate everything wuz double – two bosuns, two wireless operators, two cooks, the lot. That meant you always got yuh rest when you wun't workin'. You used t' do six hours on an' then you used t' go below f' six. An' that period o' rest wuz always guaranteed you. The skipper an' mate used t' change watches with each other, an' the engineers used t' work theirs among theirselves. They dint even use t' come on deck, the engineers dint.

'Soon as you'd finished yuh spell on deck, you used t' git a drop o' rum or whisky or gin – whatever yuh liked. You used t' carry bond on board. Yis, that yuh did. You'd hev about 20 bottles o' spirits alt'gether, an' then there'd be sherry an' fags as well. When you were steamin' hoom, just afore you got inside the limit, the ow man used t' open up the bond an' you could buy what wuz left. But that wuz really put aboard because o' the cold weather. Mind yuh, that come orf yuh expenses when yuh settled. Course it did! Some o' the ships, their crews used t' drink it afore they even reached the fishin' grounds, but our ow man never believed in that. "Thass put aboard for a purpose," he used t' say. Another thing he believed in, along o' the owners, wuz good grub. You used t' git a proper breakfast on board – cereal, bacon an' eggs, whatever yuh wanted – an' there'd be proper dinners an' teas as well. Our cooks used t' make bread an' cakes an' all sorts o' things. Afore the war that used t' be fish f' breakfast, meat an' vegetables f' dinner an' fish f' tea. An' what bread you took wi' yuh hetta last the trip. Cor, that used t' be funny stuff by the end!

'We used t' be away about 25 t' 28 days on the distant-water fishin' an' we usually got about 48 hours in port when we got hoom. That used t' take us about 24 hours t' clean the ship up a-comin' hoom an' she hetta be spotless. One o' my jobs wuz t' clean the whistle an' the siren. They were both made o' brass an' that used t' take me nearly a full watch t' git 'em nice an' shiny. I wuz bosun, like I said afore, an' while we were a-fishin' I used t' hetta do all the cod end work. You know, tyin' up the knot an' untyin' it. Cor, she wuz a ship an' a half, that *Red Hackle*. She wuz about 220 foot long an' she wuz built like a destroyer. She hed them bows what used t' throw the water away from her, an' you shoulda sin what she wuz like on board! Proper beds screwed down t' the cabin floors, bathrooms, showers, the lot. She hed two radars, automatic steerin', an' the floor o' the wheelhouse wuz all mahogany woodblock. You wun't allowed t' go up there unless you were wearin' carpet slippers! No, there wun't no boots allowed on the bridge!

'The ow man used t' take an all-day watch. He used t' be on from seven in the mornin', when you hed breakfast, till seven at night, when you hed yuh tea. He used t' come on the bridge with a collar an' tie on, an' when he'd finished he used t' go down an' hev a bath. After he'd hed his bath, he used t' come up t' the bridge in his dressin' gown an' pyjamas an' say t' the mate, "All right. I'll see yuh in the mornin'." An' orf he used t' go t' bed. His berth wuz right under the bridge an' he hed a room as well where he used t' do his clerical work. I tell yuh, I wuz never on a ship like her, before or since. I dun't reckon there wuz anywhere where she couldn't have gone.

'I mean, we used t' git over towards Newfoundland in her, yuh know. Cor, there's some funny icebergs there! An' the depth o' water! We hed a thousand fathom echometer an' we couldn't touch the bottom in some places. Sometimes you hetta steam through pack-ice. You'd ease her down then till yuh got inta clear water. A lot o' the boats used t' be double-plated for'ad t' stand the bangin'. After you'd steamed through that, yuh plates used t' be like silver. Yeah, you used t' take all the paint orf. Them icebergs, though, they used t' be lovely t' look at. Oh, magnificent! They were like lovely big swans a-floatin' down an' they were all different colours − blues an' greens.

'We got a very big haul up there at Greenland once, right near an iceberg. We laid the night, an' the nexta mornin' when it come in daylight we were up agin this iceberg. I reckon that musta bin a mile long. Anyway, that wuz my watch an' when I looked in the fish-finder, all I could see wuz one black mass. We'd shot our gear, yuh see, not all that far from this iceberg, so I thought praps we'd come foul or suffin. I nipped aft up through the funnel t' go an' look at the warps − there wuz a passageway through there where we used t' leave our boots an' oilies t' dry − an' when I got there they were closed up. I phoned the ow man up from the bridge when I got back an' he come t' see what wuz up. We only hed 110 fathom o' warp out, so he thought we'd got foul gear an' he called the crew out t' haul. Cor, you oughta sin the gear when that come up. We'd just got the doors on top o' the water, when the net shot out on top just like a big balloon. That wuz like a submarine surfacin', full up wi' fish. We hed 300 kit o' sprag cod in the net an' I dun't know how many bags we dint hetta make afore we emptied it. That took us 17 hours t' git 'em all orf the deck. We hetta work two watches. Then we steamed back under the iceberg an' laid. That wuz almost as if these little ow cod were nosin' in towards that, where the water wuz really cold. The net wuz in the water only 20 minutes!

'We made a record trip that time. We got nearly 4000 kit o' fish, which wuz about as much as the boat would hold. When we were steamin' hoom, we run inta some very bad weather orf Rockall. We were runnin' afore the wind an' we got on the bank. I never thought we'd git orf. I dint think she'd survive it. We were in all this broken water, so there wuz nothin' under us an' we hed this terrific weight o' fish on board. Well, the ow man, he called f' the chief t' give her everything he'd got. "Give it to her when I ring down," he say. An' thass what the chief did. That wuz what pulled us through. That pulled her hid up, an' when she come up all you could see wuz the bridge. The two little boats aft, they were completely covered up. I never thought she wuz a-goin' t' make it. But she did.

'Course, the ow sailin' boats used t' go up t' Greenland at one time. That make yuh think, dun't it? I dun't know how they used t' do it. They used t' go up there, a-longlinin'. When I wuz up there, you used t' git these little ow ships out o' Aberdeen, a-longlinin' f' ollabuts. They used t' take a deck cargo o' coal an' a fish-room full o' coal f' the journey up there, an' then they'd coal somewhere in Greenland afore they come hoom. You used t' be able t' git coal at one or two places up there; so yuh did fuel oil. We were up there once, when the ow man called me up t' the bridge. ''Take a look through these glasses, Pud,'' he say. So I did. Then he say, ''I've bin told that wherever yuh go in this man's world, you'll always see a Low'stoft drifter.'' Well, then I saw her, the little ow *Boston Mosquito LT373*! There she wuz, plain as day. Arthur Larner wuz skipper of her, him what hed the *Warbler* at one time, an' he wuz up there after ollabuts. Yeah, they were goin' up the Davis Straits an' onta the banks after 'em.

'Another thing we used t' see up there were the ow weather ships. There used t' be two of 'em on station at Greenland. They were corvette sort o' things an' they used t' give the weather reports every day. They used t' watch the ice breakin' up as well an' keep an eye on it when that got inta the Gulf Stream. If the bits o' ice were too big, they used t' blow 'em up 'cause they were a danger t' shippin'. We used t' take the blokes on board the weather ships newspapers an' that, an' when we steamin' hoom we used t' git their letters orf 'em f' postin' in England.

'When we'd made the last haul on that distant water fishin', we used t' cut the trawl an' dump it over the side. Yeah, we never bothered t' steam hoom with it. Mind yuh, you used t' save the cod end an' the hidline floats, but the rest you used t' chop away. Yuh doors hetta be dropped an' lashed so many feet orf the deck, an' all yuh bobbins hetta be chained back along the rail so they dint roll about when you were steamin'. Then you'd go an hev a look at the fish-rooms t' make sure everything wuz all right down there. Once you'd done that, you used t' lash yuh scupper doors back on their chains, an' then you were riddy t' go. She hed four scupper doors an' you could drive a bus through 'em very near, so yuh know how big they were. Soon as you'd started steamin', you used t' lock yuh galley doors an' no one wun't allowed on deck then till yuh got hoom. That wuz too dangerous. We all used t' hev a bath afore we started, yuh know; then we'd be away.

'Greenland wun't too bad t' work really. Nor yit wuz Iceland or Jan Mayen Island. The one place I really detested wuz Bear Island. Thass one big rock an' there used t' be American scientists on there. They used t' bring 'em orf about September time. There wuz one or two little donkeys on there as well an' they were t' do with the coal mines. You used t' git gret big jumbo haddicks up there, but the weather used t' be snow an' ice nearly all the year round. We nearly hit the rock one mornin'. Yeah, that we did. That put the wind up o' us, I can tell yuh. The only thing what kept us orf wuz the broken water round it. Oh, thass a deadly place, that is. I never liked goin' there.

'One thing you hetta watch up in all them northern waters wuz icin' up, specially if the wind wuz north-east or south-east. See, an' the harder you steam, the more you collect it. Thass the reason why they made a lot o' the trawlers with tripod masts the latter part o' the time. You know, the forem'st hed the middle piece an' two supports instead o' all the riggin' you used t' git on the older boats. Course, that riggin' used t' ice up real bad. Yeah, you used t' hetta chop it orf every two or three hours durin' the winter months. The ow man'd be out there an' all, an' you used t' hev these little ow choppers like the boy scouts hev. You'd be chip, chip, chip, chip, chippin' away an' then you used t' git a drop o' rum after you'd finished. Course, you couldn't afford t' let the boat git top heavy. She might start t' list if she did an' then you know what'd happen. On the *Red Hackle* we hed hot-water hoses t' use on the ice, but on the old boats that wuz all done b' hand.

'Another place we used t' work wuz Novaya Zemlya. Thass up the top o' Russia an' there's a big plaice ground there. That wuz so cold up there that the engineers used t' be firin' up wi' coats on down in the stoke-hole. An' not only that – the ice used t' come through the side o' yuh bunk just like mildew on bread. When yuh hauled, yuh trawl used t' go as hard as a plank soon as that come up out o' the water. You used t' go around with a hot-water donkey t' thaw it out. As well as plaice, you used t' git cod an' ollabuts up there an' all. An' do yuh know what? – that used t' git so cold up there sometimes that I've sin seagulls die o' the cold. Now a lot o' people wun't believe that, but it's true. You used t' git this black frost, an' that wuz like a thick fog a-blowin' up. I dun't know what caused it, but that wuz like a ton o' bricks a-hittin' on yuh. You'd see all the riggin' start t' freeze up an' you used t' haul straight away. When you dropped yuh fish on deck they'd freeze solid. Yeah, you couldn't git a knife in 'em an' they used t' turn black. They wun't no good then. All you could do wuz wash 'em out through the scuppers.

'When you went up t' Novaya Zemlya, you used t' steam up through the Norwegian fjords. You used t' run right acrorss t' Landegode an' pick yuh two pilots up, an' you'd stop there an' coal as well if you were on a coal-burnin' boat. Once the pilots were on board, you used t' hetta do exactly as they told yuh. They were three-ringed blokes, yuh know, an' they used t' tell yuh what course t' steer, what speed t' do an' everything. Durin' the summer you'd sometimes only hev one on board, but winter time that wuz always two. After you'd picked them up, you used t' run down t' Harstad for ice an' then on t' Tromso. You'd praps call in there f' a night an' hev a drink, then you'd be away up past Hammerfest. You used t' drop yuh pilots at Honningsvag an' you were on yuh own then. They'd done their job an' you used t' steam round the North Cape then, right down t' Vardo Island. Once you got there, you used t' run orf about 200 mile t' the East Bank in Russian waters.

'You wun't inside their limit when you were fishin'. No, that you wun't! There wuz always one o' their destroyers a-messin' about, keepin' an eye on yuh. I have bin right up the Barents Sea. They used t' lay up there for yuh, waitin' for yuh t' go inside the limit. Soon as you got near, they used t' put a shot over yuh, just t' warn yuh. An' they used t' come up alongside an' tell yuh. The ow Iceland gunboats used t' keep an eye on yuh as well, when you were up there, but there wun't the trouble up there afore the war that there hev bin since. No, that wuz only a 12 mile limit then, so what you used t' do wuz wait f' the gunboat t' go, then nip inside an' git a bit o' fishin' in.

'Up in the White Sea used t' be good fishin'. You used t' git cod an' plaice, haddicks, coalies, ling, whitins, all sorts. You dint use t' git no rooker, though. No. You used t' git quite a few o' them ow tusks, but they wun't no good so you used t' hull them anywhere.

Catfish were another thing. Yeah, you used t' git the yeller ones an' the blue ones. They're a lovely fish t' eat, but they're buggers t' gut. Yeah, you git a crowd o' them an' you know all about it. They can bite a broom handle in half, they can! If we got a lot of 'em, we used t' stand in empty carbide tins t' gut. Then they couldn't git a grip on yuh boots. Sometimes we'd crack 'em acrorss the skull with a shackle afore we gutted 'em, then we'd cut the tail orf, skin 'em an' hang 'em up for 24 hours t' bleed. The blood used t' all run out an' leave the flesh nice an' white.

'There used t' be lovely sandy bottoms up on the distant water. You'd git the odd rock or two, but nothin' much. There wun't no wrecks t' worry about neither an' one trawl would usually last yuh the trip. You worked hides on the cod end, but then rubber mats come in the latter part o' the time. That wuz a good job an' all really 'cause some o' them hides were terrible things. You used t' stick 'em in a sort o' blue disinfectant stuff an' hell if they dint stink − specially when you got back inta British waters an' the sun got on 'em! One or two o' the boys got anthrax orf 'em, yuh know. Yis, that they did, orf them foreign hides. So they stopped usin' 'em then.

'When you were up there, you dint hev no one watchin' the warps. No, not after the war you dint. There wun't no need with all the echometers an' fish-finders an' things you hed. O' course, when I first went t' Fleetwood, there wun't no such things as echometers. You hetta use the hand-lead then. An' when you were workin' deepish water, like the 200 fathom odd orf Kilda, you hed a little steam lead. You used t' hev a little, tiny winch t' heave it up with. Before the echometers come in, you used t' hefta cast the lead t' find out what yuh ground wuz like an' just take a chance on where yuh were towin' to. When the Decca come in, that wuz a big improvement. Oh, they're marvellous things, they are. There's no doubt about it. Yeah, if they're looked after, they're dead accurate.

'The main thing t' watch out for on that distant-water fishin' wuz bad weather. Thass like the *Gaul H243* gittin' lorst up there orf the North Cape. They dun't know what happened to her, do they? No one never will. Some people reckoned the Russians got her, but thass a load o' rubbish. Whatever happened t' that ship happened in seconds. Either she wuz iced up an' capsized, or she wuz a-towin' afore the wind an' come tight. If she wuz a-towin' afore the wind with her gear down, she coulda come tight an' bin pulled down arse-first. From what I've sin o' them stern trawlers, they aren't built right. They're all upper structure, an' they've got the wheelhouse as far for'ad as they can git it. An' they hent got no draught neither. I wun't care t' go in one o' them.

'No, trawlin' is dangerous enough as it is. I wuz there one year in the *Red Plume LO419* when a bloke went round the winch. She wuz an oil-burnin' boat, about 140 t' 150 foot long, an' she used t' be called the *Clevela FD94*. This chap wuz just steppin' over the warp, when a splice caught hold of him an' pulled him inta the winch. That took his leg orf up above the knee an' we hetta run him inta Reykjavik. What saved his life wuz us shovin' what remained o' his leg inta a sack o' flour. We did that t' congeal the blood, an' the doctor said that that wuz what saved his life. Yeah, if we hadn't done that, he wun'ta survived. Another thing you hetta watch out for wuz blokes gittin' hands an' arms jammed in the doors. Now I know thass a thing that shouldn't ha' happened 'cause you were sposed t' wait till yuh door wuz steady. But sometimes, with a breeze o' wind, you'd hetta keep pullin' it orf the side t' git it steady. You couldn't help it.

'Cor, that used t' be wicked up there sometimes, workin' in the cold. An' yit the more clo's you hed on, the worse orf you were. You used t' sweat, an' then when you took yuh oily an' that orf you used t' git cold. I never used t' wear hardly anything. I used t' hev a jersey on an' a pair o' fear-nots underneath me oily, an' that wuz all. Some of 'em used t' wrap up wi' scarves an' that, but I never did. When I first went round t' Fleetwood, we used t' use the old-fashioned oilies, the sort what yuh hed when yuh went herrin' catchin'. When you took 'em orf, they used t' stand up alone! Well, they done away wi' them after a while an' then the rubber ones come out. Now they've got them plastic ones an' they're a good idea 'cause they dun't freeze.

'When you were up there an' the weather wuz cold, you used t' dip yuh hands in whale oil. You used t' hev a bucket o' that standin' about time you were workin'. That go t' a certain temperature, yuh see, an' that dun't freeze, so you'd rub yuh hands wi' that an' then put yuh rubber gloves back on. Some o' the blokes used t' git inta the galley an' shove their hands round the kittle or dip 'em in hot water t' git 'em warm, but the ow man used t' swear at 'em if he caught 'em doin' that 'cause there wuz a chance o' them gittin' frostbite. "Keep rubbin' 'em in that whale oil," he used t' say. An' that wuz right, 'cause although that wuz very cold that never used t' freeze. I dun't ever remember seein' anyone wi' frostbite on the face, but yuh eyelashes used t' git a good length because o' the ice on 'em. An' that used t' freeze yuh ow lugs as well. That it did. When you got that black frost a-blowin' in up there at Novaya Zemlya, you used t' haul yuh trawl an' go below. You had to. You couldn't work in it. That dint matter where yuh were, though — the one thing you all looked forward t' hearin' wuz the ow man say, "Right, boys, this is the last haul." That used t' be a melody, when yuh got that cheery word. That it did. You knew you were goin' hoom then. Finished!

'Like I say, that wuz all clear fishin' up north. The owners used t' do well. That trip we hed in the *Red Hackle,* where we got that 4000 kit, we made £12,000 when we landed. What'd it be worth t'day, do yuh reckon? It wuz a record in Fleetwood at the time. When we passed £4000 on the distant-water boats, the ow man used t' say, "Thass all right. Now we've got the expenses." See, oil wun't all that cheap, even in them days, an' nowadays thass far worse. I dun't see how half these big ships are goin' t' be able t' keep goin'. They just wun't be able t' do it. Mind yuh, there's always bin bad times in the fishin'. Things were in a poor ow way here in Low'stoft just afore I went round t' Fleetwood. Yeah, you'd hev about one Woodbine betwin 20 of yuh, an' there used t' be no end o' blokes standin' around near the market, just waitin' t' be set on. There'd be swarms o' deckies, an' then you used t' git the poor ow herrin' skippers what hent earnt no money durin' the hoom fishin'.

'Course, you could earn good money up on the distant water. I mean, I went up t' Faeroe, I went t' Iceland an' Greenland, an' I went t' Bear Island an' Novaya Zemlya. There used t' be a good bit o' competition t' git on the fish, yuh know. Yis, there were a lot o' boats fishin' up there an' if you dint watch it someone'd soon push yuh out. They'd come edgin' up to yuh so you'd gotta give way to 'em. You hed yuh gear down an' so did they, but you hetta give way accordin' t' the rules o' the road an' then you'd find yuhself orf yuh ground an' in the deep water. Yis, you'd see his red light on yuh starboard side an' you hetta git out o' the way. Soon as you were orf the bank, yuh warps used t' drop down

125

right steep. Hell if there wun't some language then! Mind yuh, we used t' do it to other boats as well. Course we did! That dint matter whether they were foreign or English.

'You used t' git all sorts fishin' up there. There'd be the Yorkies, the Hull boys; then there'd be the Grimsby lot an' us from Fleetwood. Then you'd git the foreigners. There used t' be a helluva lot o' them. Yis, there wuz Frenchmen, Germans, Spaniards, Portuguese, Dutchmen. You name 'em, they were there! The foreigners were very well organised; they used t' hev a hospital ship with 'em all the while. Now thass one thing we never had. We used t' hetta rely on the foreigners. They'd help yuh out if yuh needed it. Oh yis, they were very good; they'd send a doctor aboard t' see what wuz wrong. Course, the skipper hetta pass out in first aid an' there used t' be a big first aid box kept in his berth, but there were certain things he couldn't do or dint know about. He used t' be able t' lance a finger or splint a leg, an' he hetta know about drugs an' poisons as well, but that wuz as far as it went.

'You used t' fish some rum places up there, yuh know. There's one called Andenes at the top o' Norway. You could tow down the tide all right, but you couldn't go up aginst it. There wuz such a strong current runnin' that you'd lose all yuh gear if you tried t' tow up it. The tide used t' run south-west there an' you used t' bang yuh gear over the side an' go down with it. You always got more fish goin' down the tide anyway, so that wuz all right. Ow cod are always on the swim an' they're always hungry. We used t' bag all the cod roes an' milches on board an' put them t' one side. They used t' fetch a mint o' money when yuh got hoom, though you dint see much of it 'cause they were sold as part o' the trip.

'That dint matter where yuh went up in them northern waters, all the fish were biguns. I mean, them plaice we used t' git up at Novaya Zemlaya, one o' them wuz big enough t' feed a fam'ly. Course, you got the ow slabs after they'd spawned, but you used t' hull them away. So yuh did the slink cod. You used t' dump all them. Do yuh know, I've bin up there an' sin three cod to a ten stone box! They were gret ow biguns, donkey's years old. The ow man used t' say, "Look at 'em an' see if they've got whiskers on, Pud." You used t' git some beautiful haddicks an' all. Yeah, you used t' stand 'em all up on their stomachs when yuh shelfed 'em an' they used t' shine like silver. I've sin some smashin' ollabuts too. I've sin some up there 36 an' 37 stone. They were about a foot thick through the neck. Mind yuh, they're a fish I've never loved, ollabuts. I wun't say thank yuh for 'em. That all depend on yuh taste, I spose. That wun't do if we all liked the same thing, would it?'

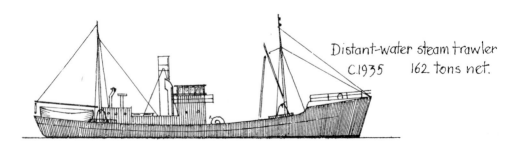

Distant-water steam trawler
C.1935   162 tons net.

The long warps have been hauled in by winch, the net heaved over the side by hand and the cod end hoisted inboard on the gilson. The bosun tugs at the cod end knot to release the bulging catch on deck. The trawl will be shot again before the crew begin gutting and stowing this haul. That task will probably last until it is time to haul again. On the grounds round-the-clock work was the common lot of trawlermen.

A smack's trip of fish laid out for sale on the Lowestoft market round about 1907. The trunks in the first row contain roker (rays), and the complete market personnel appear to have assembled for the photo. Ben Mummery whose name appears on the board above was the most famous of Lowestoft fish merchant in the Edwardian period. He is probably the commanding figure in white shirt and bowler hat.

CHAPTER TEN

On The Market

The Trawl-fish Trade

'The world is full of trouble from the cradle to the grave
To those who have to live on shore and those that plough the wave.
But fishermen both one and all have trouble night and day,
And everybody in the land will hold with what I say.'
(Anon – 'The Fishermen's Strike')

Once the last haul was made and the steam home completed, the fish had to be landed and sold. Although there were variations from port to port, the basic procedure was the same: catches were unloaded from the boats and sold by auction directly on the quayside. The method still holds good, though in the case of Lowestoft most of the species sold now have a reserve price. For example, at the time of writing, the price of a ten stone kit of plaice is £30 and a lot of resentment has been caused among fishermen at seeing Dutch plaice bought by local merchants for £8 a kit less. This has given rise to all sorts of speculation about the Dutch fishing industry being given 'hidden' subsidies, but one has to consider that very little plaice is eaten on the continent, and since a lot of plaice is caught by Dutch beam-trawlers in their hunt for soles, then why not sell the unwanted fish in a port where they have always been in great demand? And why not undercut the local reserve while you're at it?

Fishing has always been an industry of economic ups and downs, and Ernie Armes (born 1902) recalls what things used to be like on the Lowestoft market, where he spent practically the whole of his working life.

'When a trawler come in, the crew used t' unload the fish. The mate'd be ashore, watchin'; there'd be two blokes on deck, workin' the pulley an' the ropes; an' the others used t' be down below, puttin' the fish inta baskets. When the two men on the pulley pulled the basket up out o' the hold, another one used t' git hold of it an' walk it t' the side o' the boat. Soon as he put it on the rail, the lumper used t' stick his hook in it an' bring it ashore. There were two firms down on the market, Con Harvey an' the Gale Brothers, an' they employed the lumpers. They hed about three or four blokes each an' they were the ones what used t' run up the plank, hook the basket o' fish orf the rail, bring it ashore an' tip it out on the sortin' board. They were rum ow boys, some o' them lumpers – ow Dan Hitter an' Sharky Barnard an' them. When there wuz a high tide an' you got a big-bowed boat in, that used t' stand miles up in the air. What they used t' do then wuz go halfway up the plank an' wait for the bloke on board t' push the basket down to 'em. Then they'd hook it an' bring it ashore. That ow plank used t' git very slippery an' I've sin 'em fall in the harbour many a time. Yis, they'd fall in the drink! Once the boat wuz unloaded, they'd pull the planks ashore an' wait f' the next one. They got paid on what they landed, so much a box or so much a kit, an' they earnt good money. Mind yuh, they did earn it – them baskets hed four or five stone o' fish in.

'Once the fish wuz on the board, that all had t' be sorted out inta various sizes. There wuz about seven or eight different sizes in plaice, half-a-dozen in haddicks, five in cod, six or seven in soles. An' thass how you used t' go on. After the fish wuz sorted, that all hetta be packed inta wooden trunks or kits. When I first started down on the market, that wuz all trunks − wooden boxes what held about five stone − but then a bit later on the kits come in. There wuz ten stone in a kit. At one time all the cod used t' be laid out on the floor o' the market in scores, twenties, accordin' t' size, an' some o' the old buyers knew exac'ly what weight wuz there just by lookin' at 'em. People like old man Arthur Evans an' his foreman George Barbor. They'd look at a row o' cod an' weigh it up in their minds. "Eighteen stone," they'd say, an' when you took 'em out an' weighed 'em on the scale, they weren't far out. They were clever people.

'Course, I worked f' Arthur Evans f' quite a while, an' a damn good man he wuz. We used t' git paid accordin' t' what sort o' boat we packed. If that wuz a smack, you got 4/-. If you packed a drifter-trawler, that wuz 6/-. If you packed a bigger trawler, like a Crown boat, you got 7/6d. That wuz irrespective o' what quantity o' fish you landed. If you got an extraordinarily big trip, you might git given an extra shillin', but that all depended on yuh employer. Some were all right. Most were bloody tight! The watchman, who used t' watch a boat layin' at the market, he used t' git four bob a night. Four shillins f' watchin' a ship worth I dun't know how much an' full o' fish inta the bargain!

'There were sev'ral firms down on the market who the packers worked for. There wuz Arthur Evans, Cutty Robbens, Billy Williams, the Consolidated, James Sladden − yis, an' a whole lot more. Each firm had their one foreman packer an' another man along o' him. Most o' the other ones used t' be set on casual 'cause there wuz always a good many people hangin' round the market, waitin' t' earn a few coppers. As soon as the trunks an' kits were empty, the foreman used t' collect 'em up, wash 'em out an' stow 'em up riddy f' next time. After he'd done that, he dint hev much t' do f' the rest o' the day, but he hetta be down on the market agin the next mornin' at three o' clock. The rest o' the packers used t' start at four − earlier if that wuz a very big trip − an' pack the fish riddy f' salesman t' sell the fish t' the buyers. If you were a good packer, you'd make a trunk or a kit look full, but there wuz a hole in the middle where there wun't no fish. Oh, there wuz all the tricks o' the trade goin', but the buyers were wise t' most of 'em. That they were.

'There used t' be a little swindle what the buyers used t' work with the salesmen. It went like this. Say a trip o' plaice wuz bein' sold − well, someone would buy so many kits at £2-4-0d, we'll say f' argument's sake. The bids went in shillins, yuh see. Well, after a bit, one o' the buyers would shout out, "Half an' half! Four kits half an' half." That meant he wuz splittin' the shillin' bid, so instead o' payin' £2-4-0d a kit, he'd git 'em f' £2-3-6d. Sometimes, if he bought a large quantity half an' half, more'n he wanted, he sell some at the full price an' make sixpence a kit profit. That wuz what they called second-hand buyin'. Another way the buyers used t' gain wuz if they paid their bills t' the salesman by Friday, they used t' git a discount − I think that wuz ten per cent − orf what they owed.

'Course, the blokes what done all the hard work never got much! No, you dint even hev a place t' hang yuh coat up. You hetta hang it on any nail you could find. An' you dint git paid no overtime neither. That all depended on the guvnor in them days. If he wuz a decent sort o' bloke, he'd say, "You've done a good week's work. Here's a couple o' bob extra. Treat yuhselves to a drink.'' I mean, I've bin down on that market quite orften at weekends icin' fish what hent bin sold. You used t' pack 'em in big boxes an' ice 'em. You might be down there till six or seven o' clock in the evenin'. You dint git nothin' for it. While the work wuz there, you hetta do it an' no argument. You wun't in no unions then! I hev bin down there on Sunday mornins icin' back − you know, puttin' fresh ice on the fish − an' all for about a couple o' quid or fifty bob a week! But the point wuz this − in them days there wuz always about ten blokes lookin' out f' one man's job.

'That wun't just the blokes on the market what hed it tough either. The fishermen did as well. Poor ow buggers! A lot o' the smacks used t' lay on the harbour beach in the Waveney Dock. The crews used t' pull 'em up on the dolphins as far as they could go on high water; then when the tide went down, the boat used t' heel over. You used t' see the blokes scrubbin' away like the dickens, gittin' all the weed an' muck orf the bottom. Then they used t' give the boat a quick tar over afore the tide come up agin an' she wuz afloat. Cor, they dint half use t' work on them ow smacks! Sometimes they dint berth till about ten o' clock at night, but they'd be down at five the next mornin' t' git the fish out, ice the boat, coal her an' git everything riddy t' go t' sea agin. Then praps they'd be all riddy t' go an' the skipper would be stuck in the "Bank Stores". The crew'd be waitin' on board with the mains'l an' mizzen set, but with the boat tied up. About three o' clock time, away the skipper used t' come, drunk as hell, an' out they used t' go.

'Then you hed the steam trawlers. The Consolidated hed a big fleet o' them in Low'stoft; but if they dint make about £120 a trip, they used t' lay 'em up for a bit. Everybody on board got the sack an' they used t' send 'em up-through-bridge an' lay 'em up in The Crick. Praps they'd hev 20 boats at a time layin' up there. Then ow George Frusher, the manager, would say t' Herbert Breach, the ship's husband, "I dun't know, Herbie. I think we'll give ow so-and-so a start. He can hev the *Ipswich LT128*. Tell him t' go an' ship his crew." Course, there used t' be loads o' skippers an' mates what were out o' work standin' around in Waveney Road. Where you turn the corner out o' Waveney Road, goin' towards the bridge, that used t' be called "The Skinners' Knoll". See, all these out o' work skippers an' mates used t' hang round there, waitin' for their pals what'd got a job t' take 'em inta the "Bank Stores" or the "Anchor" an' buy 'em a pint arter they'd landed. If they'd made a trip that wuz! Yeah, they used t' be waitin' there t' skin 'em for a drink. They were glad t' git anything in them days. I should think nearly every other man in Low'stoft wuz out o' work betwin the wars, an' there wuz no sympathy from the boat-owners. They were just like the farmers. Boat-owners an' farmers never hed any sympathy f' their workers, an' yit if anything happened t' them they wanted the gover'ment t' help 'em!

'Do yuh know − the Low'stoft market dint close down durin' the General Strike. That dint make hardly any diff'rence to it at all. When the trains stopped runnin', the merchants chartered a boat, loaded it up wi' trawl-fish an' sent it up t' London. They used a lot o' the local haulage firms as well. The lorries used t' run the fish t' Billin'sgate. They hed a notice on the back which said "FOOD" an' they used t' git through all right. There wun't much trouble wi' pickets, I dun't think. An' there wun't much trouble on the market here 'cause in them days there wuz very few people in the fish trade what were in unions. That wun't organised at all. The boats managed t' keep fishin' 'cause there wuz a fair bit o' coal in the town, an' when they run short they used t' bring coal bricks acrorss from Holland an' Belgium. The Consolidated sent one or two o' their ow trawlers over. The *King Canute GY1124* wuz one, an' then there wuz a little ow coaster-thing called the *George Frusher.*

'That used t' be damn hard work down on the market at one time. I mean, the floor wun't all nice level concrete like it is now. It wuz a series o' big granite slabs, with cracks in between, an' you'd be pushin' a barrer along an' you'd go an' git a wheel jammed in one o' the cracks. Over yuh fish used t' go! They were just an ordinary sack-barrer what yuh used an' they used t' take some pushin' over them slabs. You dint hev no protective clothin' in them days neither. Nowadays the market workers are all rigged up wi' oilskins, rubber boots an' rubber gloves, but in the old days all you hed wuz a pair o' woollen mittens in the winter-time. Every now an' agin you used t' pull 'em orf an' wring the slime an' muck out. Then you'd put 'em back on agin.

'At one time there wuz no filletin' done down on the Low'stoft market. No, none at all. That only started when one or two o' the Grimsby firms come here. That wuz a whole fish trade in Low'stoft. Prime stuff an' plaice. Yeah, this wuz a prime port. Grimsby wuz a bloomin' shit port! Thass why they went f' filletin'. You can hide things up that way. Once that'd started, the Low'stoft people hetta go into it t' compete. That made things cheaper, yuh see. I mean, every fish shop used t' employ a blockman at one time just t' clean the fish whenever a customer come in. Well, when a fishmonger could buy fish that wuz already filleted, that done away with the blockman. There used t' be all sorts o' ways to prepare fish at one time. They used t' curl whitins an' crimp haddicks, for instance. Both o' them were similar things. The fish used t' be skinned, the backbone taken out, the head left on an' the whole thing bent round so that'd lay nicely in the pan. You never see that now. Everything is filleted. You mustn't hev fish wi' bones in nowadays!

'There used t' be all sorts o' stuff handled down on the market at one time. Plaice, soles, turbot, brill, whitins, haddicks, gurnards, weevers, rooker, cod. You never see gurnards or weevers landed there now, an' they're lovely fish t' eat. So are latchets. They're like a big gurnard an' they're lovely when they're baked. Mind yuh, you git a lot more monks an' catfish now 'cause the boats go further down than what they used to. The further nor'ad you go, the more you git o' them. They used t' land one or two sturgeons in here at one time. Do yuh know, there wuz a sturgeon landed in at Low'stoft just after the last war an' the crew hed bin an' gutted it! Bang went the caviar! I can remember one o' Arthur Evans's smacks bringin' one in afore the war an' that went orf t' London. If I remember rightly, Sam Isaacs bought it.

'You used t' git other things as well. There'd be the odd boat with a kit or two o' conger. An' praps you'd git a few tusks an' coalies if the boats hed bin down north. There wuz a lot o' dorgfish o' course — well, dorgs an' nurses. They're a similar kind o' fish, but they aren't exac'ly the same. Most o' the rough stuff went t' the fryin' trade an' the small fish shops. You used t' git a lot o' the little Norwich buyers down on the market at one time an' they used t' take the fish back with 'em, skin it an' clean it in their shops, keep some, an' supply other shops with the rest. Sometimes they'd git orders for other people an' drop it orf on the way back t' Norwich.

'There used t' be several foreigners land at Low'stoft at one time o' day. The Oostend men used t' come here a good bit, so did some o' the Dutchmen an' one or two French. If they were fishin' orf here, they quite orften used t' run in rather'n take the fish back hoom. Course, there weren't many restrictions afore the war an' the packers liked t' see 'em come in 'cause they could fiddle the foreigners better than what they could their own boats. Ow Cutty Robbens used t' sell for the Belgians an' them, when they come in, an' they used t' be quite a sight in the port. One thing they used t' do wuz put a string through the gills o' their small plaice an' dabs an' hang 'em from the mizzenm'st t' the stern o' the boat so they dried in the wind. An' they used t' eat 'em like that. Yis, that they did, but there ent none o' that go on now.

'At one time o' day a couple o' blokes called Alger an' Bird hed permission t' collect all the refuse up orf the market. They kept it clear o' all the fish heads an' skins an' bones an' that sort o' thing, because once filletin' started you hed a lot o' that. They used t' come round with a barrer an' shovel up all the muck orf the market, stick it in their lorry an' cart it round t' the various farmers. That all used t' git ploughed inta the land f' manure. I've sin Carver Hadenham from Grove Farm there at Pakefield come down the market with a gret big ow tumbril an' take away a load o' fish what were too small t' eat. I dun't know whether he paid for 'em or not, but he used t' cart 'em away an' plough 'em in. Just after the war, about 1948 or 49, the Grimsberians come here – a proper manure firm – an' they got permission t' hev a railway truck stand at the back o' the market for all the muck t' go in. They used t' buy it from the merchants, where Alger an' Bird got it f' noot. Well, o' course, that done Alger an' Bird out o' their business. All the stuff put inta this truck used t' go away to a fish-meal factory an' that wuz a good source o' income t' the merchants t' be able t' sell their waste. I mean, if you got five stone o' fillets out of a ten stone kit o' plaice, you were cuttin' well. But that still left five stone o' hids an' bones an' stuff. Well, if you could sell that, all well an' good. That wuz known as the gut trade, though the fish were actually gutted on board the trawlers o' course, an' a lot o' these merchants used t' do well out o' their gut cheque at the end o' the month.

'As things stand at the moment, that look as if fishin' in Low'stoft is on the down agin, or will be shortly. I ent surprised. You can't go on scoopin' up fish like they do t'day an' not feel the effect. I mean, there's everyone bangin' away as hard as they can go in the North Sea – British, Germans, Dutchmen, Frenchies, the lot. They're cleanin' it out. The boats that used t' land in here at one time were spread all over the North Sea. Not no more they aren't. Not the big ones. They're all down on the Fladen Bank, orf Aberdeen; or over on the Klondyke, orf Norway. An' when they've cleaned that out, where are they goin' t' go then? They cleaned the Dorgger Bank out years ago. You go an' shoot a net there now, all you'll git is a wreck. Thass like the Clay Deeps an' all them grounds out here – the ow smacks used t' git some nice trips from there, but these modern beam-trawlers ha' gone an' ploughed 'em up t' buggery. I dun't know how far I'm right in what I say; all I know is that you can't keep goin' t' the well for ever.'

There speaks the market-hand. The last word comes from Arthur Evans (born 1901), son of the man for whom Ernie Armes once worked and himself a fish merchant of long standing. His knowledge of the fish trade generally is rivalled by no man living:

'My father came up t' Low'stoft in about 1899. He'd been a porter on Billingsgate Market, then he went as a salesman with J. T. Clark. It was them who sent him t' Low'stoft. I think he was about the first Billingsgate salesman to move up here, so you can imagine the reception he got! After a while, he went into business on his own; and when I left school in 1916, I went in with him. We lost eight of our smacks in the First War by enemy action. The submarines used t' come up, let the crew get off, stick a bomb aboard and just blow the boats up.

'After the war a number of big steam trawlers came into Low'stoft. We ourselves had two big boats, which we bought from Aberdeen, the *Fort Edward A180* and the *Loch Loyal A132*. About two years later we'd lost over £10,000 on those two boats. They turned out to be too big for the port. We weren't the only ones t' lose money; lots of people did. I think we took these boats in about 1919 or 1920 an' two years later we'd got rid of them. When I first started on the market, I learnt the hard way. My father was a good man, but his idea of teachin' yuh to swim was t' take yuh to the deep end an' throw you in. I started by doing what all the labourers were doing. I was packing fish, running up fish an' all the rest of it. Then, as time went by I gradually got into the buying side.

'When I was a youngster there were about 20 different salesmen on the market. Now there are about half-a-dozen. They used t' start selling the trawl-fish about 7 am, or praps 7.30, an' there'd be at least five or six sales all at once. Everybody would be buying up and down the whole o' the market, and if you wanted t' get in one more'n one sale you had to run like glory from one place to the next. The only time the market closed was Christmas Day an' Good Friday. Well, there was no Sunday market either, but prior t' 1900 there was. Ow George Barbor, our manager, used t' tell me about when he was a young man, how the market used t' be open every day o' the year except Christmas Day.

'The herring season didn't make any difference to us. We carried on the business on the old Trawl Market durin' the autumn, and as soon as the home fishing was over we moved into the Waveney Dock. The trawlers all used t' land together, but they all got in without too much trouble. There wasn't half the pushin' an' shovin' there was with the steam drifters. The salesmen were either employed by the fish-selling companies or were individuals workin' on their own. One or two o' the boatowners used t' have their own salesmen as well, because that way they didn't have to pay out any selling commission to another firm. Old Effie Moore, the smack-owner, used t' sell all his own fish. He was in business with his two sons and they used t' do all their own selling and packing. I used t' be very pally with Billy Moore, Effie's boy, an' he used t' be one o' the best salesmen down on the market – and one o' the smartest. Always well turned out.

'When it came to the actual selling, there used t' be a senior salesman who sold the prime stuff an' the plaice, and another salesman with him who sold the rough stuff an' long fish. In my lifetime the bidding was always up, but many, many years back it was Dutch auction. Nowadays, things are very unhealthy down on the market, in my opinion. There's no such thing as an open market any more. The price is fixed, and no matter what happens you can't buy or sell fish below that price. And another thing – you can stand around a sale down on the market now, and unless you know what's actually going on you don't really know what the fish is making per kit.

'During the 1920s, when my father was going full belt, we employed about 20 t' 22 men. We were handling practically everything in those days, but our main line was soles, turbots, brills an' plaice. We never had a real trade for cod even; prime an' plaice was our big thing. It's been the same with me since the war, though of course I do deal in other fish. I mean, I'll handle whitings an' lemon soles, but I don't touch haddocks much nowadays because they're too unpredictable. You might see 50 or 60 kit landed one day, then go for a week without seeing hardly any. In my kind of business you need a continuity. It's no good quoting a man one day and saying, "I haven't got them" the next. He'll go where he can get them regularly. In the old days we used t' buy a colossal quantity o' haddocks. You used to get these little Danish seine-net boats coming into Low'stoft, as well as all the local trawlers. We'd buy about 200 kits o' haddocks from them. I've seen the time when we sent away 200 14 stone boxes a day just for London, and we were only one of about five firms doing it! It suddenly collapsed of course. You can't go on fishin' like that. They gradually scooped the Dogger Bank out, fished it absolutely dry.

'A lot of the trouble in the 1930s in the fish trade was caused by over-fishing. The whole North Sea was getting fished out. See, the boats had a restricted range then, much more so than the diesel boats now, an' that meant they didn't go much beyond the Dogger Bank. There used t' be a lot of small stuff landed, and if it didn't sell it used t' get dumped. So did the stuff which wasn't very edible; nowadays they fillet it, process it somehow and freeze it. You couldn't do that in the days before refrigeration. The fish-friers used t' buy a lot o' the small fish down on the market, but we never had a fry trade because we were mainly concerned with selling prime fish an' plaice to retail outlets. I wish we had had a fry trade; that might have kept us goin' when the crash came.

'My father went out o' business more or less overnight. The trawl-fish industry was in a very bad way, an' not only that — he lost a lot o' money elsewhere as well. He had money invested in stocks an' shares, and in different companies, and he lost heavily there. As soon as we'd gone bang, I went down onta the market an' worked for the German export companies, loading klondyke herrin' aboard their freighters. This'd be about 1936. After a year or two o' that, I got a driving job with the Co-op, delivering stuff from the shop, an' then when war broke out I was conscripted inta the fire service an' went t' London. But that's another story; that's not fish. I got back inta the fish trade after the war with the help of a man named Tommy Colbeck and I've been going ever since.

'I was always dead keen t' go in with my father an' I was down in Padstow with him in the April o' 1916, when Low'stoft was bombarded by the German fleet. After the war I used t' go down on my own t' buy soles there because, as you know, there used t' be a lot o' Low'stoft an' Yarmouth boats go round just for the sole fishin'. The early months o' the year were a slack time in Low'stoft so I didn't mind the trip t' Padstow. And another thing — I was always an open air lad; I'd much rather be out than in. Obviously I had t' do a certain amount of office work, but I never really liked it that much. Our busiest time was the summer an' autumn, because by the time you got back from Padstow, in May, the plaice out here in the North Sea were improving in quality after spawning — what we called making up. As the warm weather came in, so you'd see the plaice fatten up with the warmth. It was almost the same as the flowers blooming.

'The summer was the time for soles as well. June, July an' August, they'd be beautiful; then they'd drop away a bit an' come on again December/January time. In the old days we bought more soles than anybody else on the market, I should think, an' it was the same with turbots an' brill. I don't buy turbots now. They're all goin' to the continent. Yeah, I should think 80% o' the turbots landed at Low'stoft go t' the continent. There used t' be a well-known skipper here in Low'stoft at one time called Oscar Pipes, who used t' get my father two trips o' turbots every year from somewhere down orf Ramsgate. He knew just where t' go, an' when, but there wuz only the two trips. They were the most handsome turbots I've ever seen, I think. They were perfect. He used t' say t' my father, "Guvnor, you'll be wantin' some turbots. I'm goin' t' get yuh some. But after the second trip, even if you offer me £1000, I shan't be able t' get yuh another fish." I remember one year he came in with about 32 trunks on the first trip an' about 18 on the second. We paid about 2/4d a pound for 'em, which was a very big price in them days.

'Before the last war the prime fish used t' be handled very, very carefully down on the market. In fact, some o' the ow boys used t' get downright vicious about it. There used t' be hessian mats put down for the soles an' turbots t' be shot out on, an' the fish were handled as if they were pieces o' Dresden china. If one o' those turbots slipped off the mat onta the market floor, the man responsible would stand a good chance o' gettin' knocked flat by the skipper or the mate, who was standing by. Everything was handled perfectly. It had been from the time it was caught, what with all the washing an' everything. Nowadays they just don't bother; the fish is handled just anyhow. Also, I would say undoubtedly that the gear they use today damages the fish. Mind you, I'm not a fisherman. I'm only speaking as a merchant, so I could be wrong, but I would say that these stern trawlers are responsible for chafed fish. You never used t' see chafed fish like you do now. They're just pulled up the ramp as they come in an' that must be partly responsible. Chafed fish doesn't mean it's bad fish – except from a presentation point of view; it doesn't look good. Plaice always chafe pretty badly if they're handled roughly, but in the old days o' the smacks an' the drifter-trawlers more care used t' be taken.

'Back before the war it would be easier for me t' tell you where our fish didn't go over the country. We covered the whole country practically an' we even sold soles in Scotland, because they don't get 'em up there. We used t' send stuff as far as Newcastle, across t' Liverpool, Leeds, Sheffield, Wales, down t' Penzance, down t' the South Coast. A terrific lot down t' the South Coast, right the way along. And a terrific lot t' Billingsgate o' course. Yes, and Grimsby. We had very big customer up there for turbots. A lot of our trade was to retail (it still is) an' we used t' do our selling mainly by telegram at one time. Then, as time went on, more an' more business was done over the phone. At one time we used t' send whole batches o' telegrams away, a hundred at a time, an' the Post Office had a special place down on the market to handle it all.

'We used mostly rail for transport back in the old days. Mostly rail. Nowadays it's all done by road. We had lorries before the war, but only for running stuff from the market t' the railway sidings. I told yuh how we used t' send these big boxes o' haddocks down t' Billingsgate, didn't I? We used t' pack 14 stone into a 12 stone box. We got the boxes at 4d each from the German klondyke firms; used t' buy about 10,000 of 'em after they'd been used and clean them out an' do them up. The smell wasn't too good, I can tell yuh, when they'd had maisy herrin' in! Nowadays all our fish is despatched in cardboard boxes. Our biggest one holds about four stone, but most of our business is done in the one stone size.

'As well as lorries, there used t' be a lot of horse an' carts down on the market. In fact, in the First War it was all horse an' carts because what lorries there were had been commandeered. Durin' the General Strike I went down t' Billingsgate one day as relief driver on a three-ton lorry with about five ton o' fish aboard! We had "FOOD" up on the side o' the lorry, an' just as we got t' the other side o' Chelmsford this car comin' up the other way stopped an' waved us down. The driver was a Billingsgate carter by the name o' Race and he told us not t' go the way we were heading, but to turn off and go via Epping. This was because people were out in the streets in the East End, up-ending cars and all sorts. However, we wanted t' get t' Billingsgate as soon as possible, so we decided t' carry on. When we got t' Poplar, a police type jumped aboard just as we were asking the way an' said he would take us through. When we finally got there, what a carry-on! The blokes that were picketing the market shouldn't have been picketing there at all; they were supposed t' be picketing the boats in the river, making sure that nothing went out. Most of 'em were drunk as well. One of 'em had a real go at me, because I was prepared t' argue about bein' allowed through, what with "FOOD" being written up on the side o' the lorry an' everything. In the end a foreman porter came along, pulled a revolver out of his pocket and smashed this chap in the face with it. They all dropped back then. Do yuh know who the man with the revolver turned out t' be when we asked about him afterwards? – Seaman Harry Russell, cruiserweight boxin' champion o' the Navy!

'I didn't do many runs down t' Billingsgate, and as far as I can remember the strike didn't affect the fish trade all that much – apart from there being no trains for a bit. O' course, the railway people used t' run the market, yuh know. The Great Eastern first of all, then the L.N.E.R. They did it very efficiently. Yes, things were well run. It was the cleanest market in England, I suppose. It was scrubbed down every night – and really scrubbed down! They never touch it now, except with a hose. In the old days it was scrubbed down every night with salt water an' the place was spotless. And the service was good too. I mean, the railway workers who used t' load the fish-trains really used t' work. They'd sweat their insides out on that job, an' all for about £2 a week. In fact, I don't know whether they got £2 a week quite. But they'd do anything for yuh. Nothing was too much trouble for 'em. They were real sloggers.

'My father an' me used t' get down t' the market about 7 am. He'd get me up somewhere between five an' six, I suppose. He was good at getting up in the morning, my father. Yeah, he'd been used t' getting up at three when he was a porter on Billingsgate! During the First War the hours weren't too bad because there was only a certain amount of fish being landed, owing to the trouble, and we invariably finished by midday. Directly the war ended the hours began to increase, and we could finish at any time up to about ten o' clock at night, depending on what was going on. I should think the average wage was about 50 bob during the 1920s. I don't know if anyone was ever paid overtime, but I don't think so. Certainly there were no overtime rates. It's diabolical when you think about it now. Our star man got about £4 a week an' he was getting on for about double the others, but he was a very skilled worker. I didn't get any better payment than anybody else, I might tell you! My father used t' say, "You'll get more when you're worth it." But his idea of my value wasn't all that high for a year or two. He wouldn't let me get any big ideas above my station.

'I should think the biggest change in the fish trade in my time was when filleting came in down on the market. Yes, once the idea of filleting caught on, things changed in the trade. O' course, they'd been filleting fish for a long, long while in the shops, but nobody thought t' do it on the market. Lowestoft was always a whole fish trade; an on-the-bone trade, if yuh like. Before the Second World War there wasn't very much filleting, but afterwards there was very little else. It helped transport costs as much as anything, because six stone o' fish became three stone o' fillets an' that was a big saving. Nowadays it's all filleting an' processing, with all the leftovers going for fish meal. There's a lot o' money in that o' course, whereas at one time you had somebody pick up all the waste for nothing. They kept everywhere clean an' didn't pay a penny for what they got. With the advent of filleting there was a lot more waste an' that's when the Grimsby Fish Meal people came in. The trade is worth thousands o' pounds a year now, though it does have its ups and downs. Sometimes it'll go right down. Then there'll be a world-wide shortage o' fish meal an' up it'll go again.

'One o' the biggest problems down on the market at one time was ice. Before they built the present ice factory, the ice used t' come down from the old place in wooden kits, already crushed. The trouble was it used t' run away so quickly. You know, melt. The ice comp'ny people always blamed the water. I don't know what the matter was, but it certainly used t' run away. Going way, way back, the ice came down onta the market in slabs from the old ice-house near the bridge an' one o' the first jobs the market workers had was t' crush it all up. They used t' bring it over in ships from the Norwegian fjords; and they also used to cut it out of Oulton broad during the winter, when everything froze up. I've heard ow George Barbor say that they had t' be down on the market at five o' clock in the mornin'. He said he used t' call at the "Clapham Hotel" on the way, if the weather was cold, t' get a rum an' coffee for three ha'pence. What about that? What's the price o' coffee now, never mind the rum! Another thing about the ice was that sometimes you had t' wait before you got what you'd ordered. I don't suppose there was an actual shortage, but you had t' wait. I think that, in my lifetime, most o' the fights I've seen down on the market have been caused by ice. Two fellers would put their hands on the same kit. "Thass mine." "No, thass mine!" An' there's blood for supper in no time. You ask Ernie Armes. He'll tell yuh.'

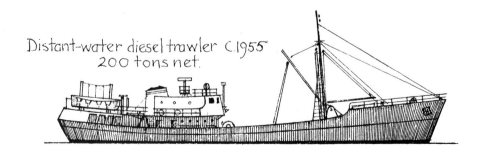

Distant-water diesel trawler c.1955
200 tons net.

Trawl fish being shot on to the sorting board. Once a boat's catch had been sorted out into the various species it was laid out for sale by auction. Too much ice made it difficult for the sorters to handle.

Last Haul, Boys

'Oh, it's three-score and ten boys and men
Were lost from Grimsby town.
From Yarmouth down to Scarborough
Many hundreds more were drowned.
In herring boats and trawlers
And fishing smacks as well
They had to fight the bitter night
And battle with the swell.'
(Traditional – 'Three Score And Ten')

The chorus from the folk song above commemorates the great October gale of 1880. The lives of the men whose memories we have heard in the preceding chapters were governed by the hazards of the sea and just as much by the economic winds. They speak of both without dramatics or heroics, but in the plain, matter-of-fact way that comes from long acquaintanceship with the sea. They speak for their generation of the economic hardship that fishing communities suffered during the 1920s and 30s, and have undergone once again in recent years since Iceland successfully enforced her 200 mile limit. Fleetwood, Grimsby and Hull have been the places most affected, with the Yorkshire port having to bear the greatest privation.

It is hard to get statistics to mean anything in human terms, but Hull's plight since the Icelandic cod harvests were placed beyond the reach of her boats has been desperate. In February 1980 all but 27 of her 130 distant-water trawlers had been laid up (and only two of these were actually fishing for the home market) and her Vessel Owners Association had gone into voluntary liquidation, with huge debts outstanding – much of the total sum being unpaid landing and harbour dues. At the end of March 1980 the first landings for two months were made in this once teeming port, but the boats were foreign ones and the harbour dues were specially lowered to attract them in. As I said at the end of the first chapter, Lowestoft seemed at first to have profited from the misfortune of the distant-water ports, but no one really believed that a new boom was about to begin. Steeply rising fuel-oil costs and the importation of good quality, cheap continental fish are just two of the factors that give cause for concern in the East Anglian town.

It is impossible to predict the future of the British trawling industry, but one thing I do know. Ever since the introduction of steam engines in the 1880s there has been a dangerous kind of in-built snakes and ladders game in the trawling industry. It works like this: bigger engines and better trawls catch more and more fish for a time, but after a while catches fall because of the depletion of stocks. Then, in order to maintain fishing at previous levels, the size of the engines and the weight and spread of the gear has to in-crease – which, again, will eventually cause a further drop in catches. And so it goes on.

Against this depressing equation can anything be done to maintain a healthy trawl-fish industry? Will Britain try to declare her own 200 mile exclusive zone (irony of ironies!) and prohibit fishing by foreign vessels within its limits? Can agreement be reached with E.E.C. countries as to quotas, and can these quotas be enforced? Will there be a positive move towards making mesh sizes larger? Would it be possible to conserve species by leaving their main grounds alone during the spawning season? Are the vessel owners prepared to invest in equipping their larger boats for fishing in deeper waters off the continental shelf, and will the food-processing companies assist them by purchasing and marketing new varieties of fish? Will there be a return to smaller craft and methods of fishing less destructive of the sea bed than those which prevail at the moment? It is difficult to imagine a return to the sailing smack, except perhaps by one or two romantics, but the wind was free! Thoughts are more likely to turn to Government oil subsidies to off-set ever more crippling operating costs.

One thing is certain. Unless a solution is found we shall witness the demise of another great British industry as we have in the last two decades seen the entire disappearance of the herring fishing. And with it we shall lose a *very special* breed of men the nation can ill afford to lose. Men like those who have told their story in these pages – brave, resourceful, unassuming men. Superb seamen. Essential food producers.

STEAM DRIFTER/TRAWLER
'TOWING ALONG'

Glossary

Allotment/allowance money	weekly payment made to share fishermen and deducted from their pay-off
Bag	cod end
Bag-rope/bag-stopper	a rope that prevented the cod end swinging too far inboard
Barking	tanning sails
Batings	upper part of a trawl net, where tapering down towards the cod end occurred
Beads	small wooden bobbins on a trawl's ground-rope
Becket	loop in a rope's end, or larger loop made by throwing a rope round an object
Beef-kettle	a large cast-iron cooking pot
Belly	the bottom of a trawl net, directly beneath the batings
Bight	a loop in a rope
Billiges	bilges
Blinder	a piece of fine-mesh net laced inside the cod end
Blobs	jellyfish
Blockman	man employed by a fishmonger to prepare fish for cooking after the customer had purchased it
Blow down	drop steam pressure and extinguish boiler fires
Bobbins	wheel-like devices on a trawl net's ground-rope
Bogeyman	fisheries patrol vessel
Boom	the spar on the foot of a sail
Borsprit	bowsprit
Bosom	the curved mouth of a trawl net
Bosun/bosman	the man below mate on a trawler
Bottles	glass floats on a trawl's headline
Brickie(s)	Brixham smack or fisherman
Bridle	wire guy that joined the trawl head to the warp on a beam trawl and the trawl door to the net on an otter trawl
Brill	bad mood
Bring up	anchor offshore
Buff	large, pear-shaped, canvas or plastic float
Bunts	parts of a trawl net's wings nearest the quarters
Buts	halibuts
Butterfly	special type of dan leno
Casing	the steel housing above the engine-room
Caulk	to make waterproof the cracks where a ship's planks join
Chipping hammer	tool used to knock the rust off metal plates, chains etc. prior to painting

'LONG' FISH.

SAITHE up to 45 ins.
'COLEY'

COD - up to 40 ins.

HADDOCK up to 30 ins.

WHITING up to 25 ins.

HAKE up to 28 ins.

TORSK 'Tusk' up to 40 ins.

LING up to 30 ins.

Clean swept	term used of a sailing boat that has had its sails carried away by the wind, and is blown before it
Cleat	piece of shaped wood or iron used for fastening ropes to
Come fast/come tight	term used of a trawl net fouling an obstruction on the sea bed
Cran	a measure of 37½ Imperial gallons; 28 stones of herring by weight
Crawlers	brittle-stars
Cutch	substance used to preserve fishing nets
Daddy	a lovely ship
Dan/dahn	a kind of marker buoy
Dan leno	a device used to spread the headline from the ground-rope on an otter trawl
Dandy	wire strop used to lift the after trawl-head on a beam trawl
Dangles	metal rings linked to lengths of chain that were threaded along a trawl's ground-rope
Dill	sump beneath a marine engine
Dodge	keep a boat head to wind
Dodger	canvas shelter under which a smack's helmsman stood
Dolphins	free-standing wooden posts or piers that boats are tied up to
Donkey	a general term for any kind of engine
Drogues'l	a conical piece of canvas, mounted on a steel frame, which was used to slow down a sailing smack as it entered harbour
Dudder	shake or shudder
Dutch(men's) farts	sea urchins
Eye	loop in the end of a rope (smaller than a becket)
Fair-lead	iron or wooden guide to direct the line of travel of warps and bridles
False bellies	pieces of old net on the bottom of a cod end, which were put there to prevent chafing
Fastener	the act of coming fast, or the obstruction causing it
Fear-nots	fishermen's trousers of heavy white material
Fiddle-fish	angel fish
Fire up	stoke up a boiler
Fish-bag	net bag for washing demersal fish in by trailing alongside the boat
Fish-room	fish-hold
Flapper/flopper	net flap that prevented fish from getting out of the cod end
Foc'sle	living quarters in a boat's bow section
Fore-peak	small storage space in the bows of a fishing boat
Fore-room	fish-hold for'ad of the main one
Forestay	the main wire brace of either the mainmast or the mizzen
Gallus	the gallows frame where a trawl door hung suspended when not in use
Garf	gaff. The spar on the head of a sail

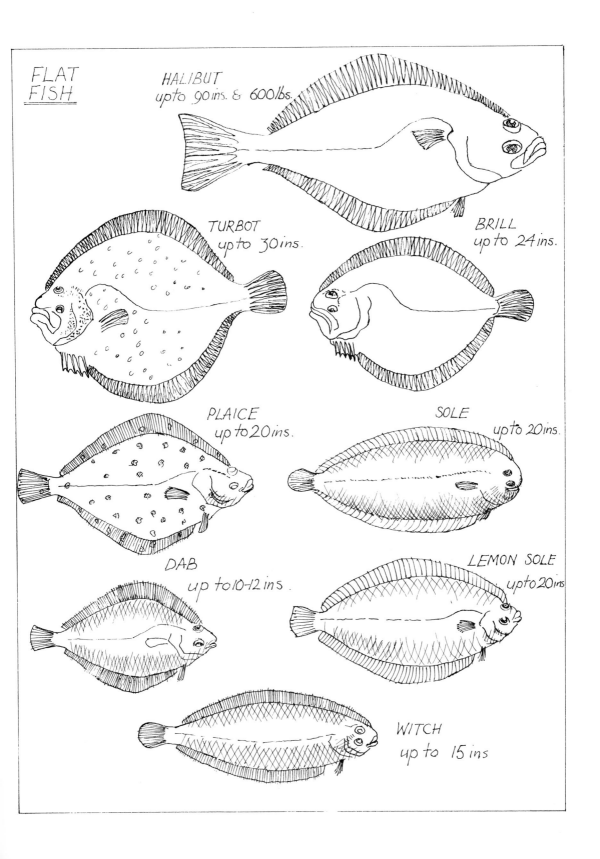

FLAT FISH

HALIBUT
up to 90 ins. & 600 lbs.

TURBOT
up to 30 ins.

BRILL
up to 24 ins.

PLAICE
up to 20 ins.

SOLE
up to 20 ins.

DAB
up to 10-12 ins.

LEMON SOLE
up to 20 ins

WITCH
up to 15 ins

Gilson	wire hoist on a trawler's foremast for hoisting the cod end
Go-ashores	fisherman's best clothes for leisure wear
Gross tonnage	nothing to do with weight. One gross ton equals 100 cubic feet of enclosed space below deck. Net tonnage is the gross figure less the boat's non-earning spaces, such as living quarters, galley etc.
Ground-rope/foot-rope	the thick bottom rope on the mouth of a trawl
Gun-layer	sailor in charge of a team of gunners
Handspike	large wooden rod used on board a smack to ease the warp along the boat's rail
Hawse-pipe	hole in a boat's bow through which the anchor chain runs
Headfish	prime fish
Headline	the top rope on the mouth of a trawl
Heaving line	rope on which a smack was warped out of dock
Hickety	rough
Home fishing	the East Anglian autumn herring season
Hoodway	the entrance to the cabin on a boat
Hook	anchor
Hovel/hubble	salvage claim for assisting a boat in trouble
Ivy leaves	small plaice
Jigger	a large rectangular topsail
Jinnies	starry rays
Jiving	a variation of gybing.
Kelly's eye	link ring on the back-strop of an otter door, connecting the door with the bridle when towing along
Kit	ten stones of fish, or the container holding this quantity
Klondyke herring	fish exported to Germany in salt and ice
Knockout	the act of releasing trawl warps from the towing block
Larnch	launch
Latchet	tub gurnard
Lazy deckie	a strop running from the cod end becket to the trawl headline. It could be slipped from the headline and hooked to the fish tackle to help haul a big bag of fish aboard
Leach-line	strop used for pulling in the ground-rope on a beam trawl
Lee tide	a tide running with the wind
Legs	(1) the plaited rope stopper that secured the warp to the towing post on board a smack; (2) wire strops between an otter door and the trawl net itself
Lengthening piece	section of net inserted between the cod end and the main part of the trawl in order to hold large catches
Light duff	dumplings
Little boat	a fishing boat's dinghy
Liver-jar	wooden barrel for storing cod livers
Long fish/long stuff	cod, haddock, whiting, haddock, ling etc.

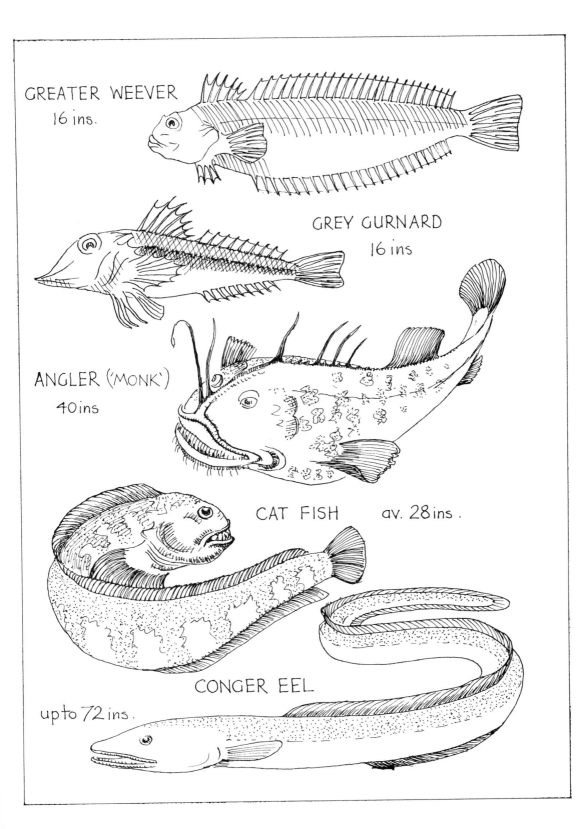

GREATER WEEVER
16 ins.

GREY GURNARD
16 ins

ANGLER ('MONK')
40 ins

CAT FISH   av. 28 ins.

CONGER EEL

up to 72 ins.

Loom	the glow of a lighthouse or lightship seen when the beam is actually below the horizon
Luff (up)	bring a boat up head to wind
Lumpers	market hands who brought fish ashore from trawlers
Maisy	term used of herring full of spawn
Manilla	vegetable fibre used in making rope
Messenger	long wire strop with a hook on one end that was used to draw trawl warps into the towing block
Milches	soft roes
Moorlog	fossil tree of the Mesolithic period, usually the lower trunk and root system
Nettle	strands of wire or twine twisted into a single length, whipped at the ends and reeved through the head of a pin to hold it in place
Oily	oilskin
Ollabuts	halibuts
Overseas	areas of turbulence where tides and currents meet
Painter	rope mooring-line fixed in the bow section of a small boat
Pair-fishing/pair-trawling	two boats working one large trawl between them
Part	break (a term used of warps)
Peds	a general term for whicker baskets
Pennant	independent strop on trawl gear that links bridle and warp, and enables continuous winching-in of the gear
Penny stamps	very small plaice
Perks	boards halfway down the hold of a drifter-trawler
Pikey/pipey	a sabellid worm, living in a tube of sand grains
Pockets	compartments on the sides of a trawl net, above the cod end, which acted as fish traps
Pound boards	planks that divided holds or deck space into compartments
Poundage	payment made to trawlermen on top of their wages
Pounds	compartments for the handling or storage of fish
Prime fish	halibut, sole, turbot and brill. The more valuable species
Pud	Fleetwood term for an East Anglian fisherman
Punch	steam through heavy seas
Quarter-rope	strop on an otter trawl that is used to haul the ground-rope in
Quarters	those areas of an otter trawl where the upper wings meet the headline and where the lower wings meet the ground-rope at either end of the bosom
Rail	top part of the gunwale
Rammy	a Ramsgate man or a Ramsgate smack
Red-noses	plaice that chafe themselves by burrowing down into the sand
Reef	to shorten sail
Reef-breeze	wind of sufficient strength to necessitate reefing
Reeve	to thread something through an eye or hole

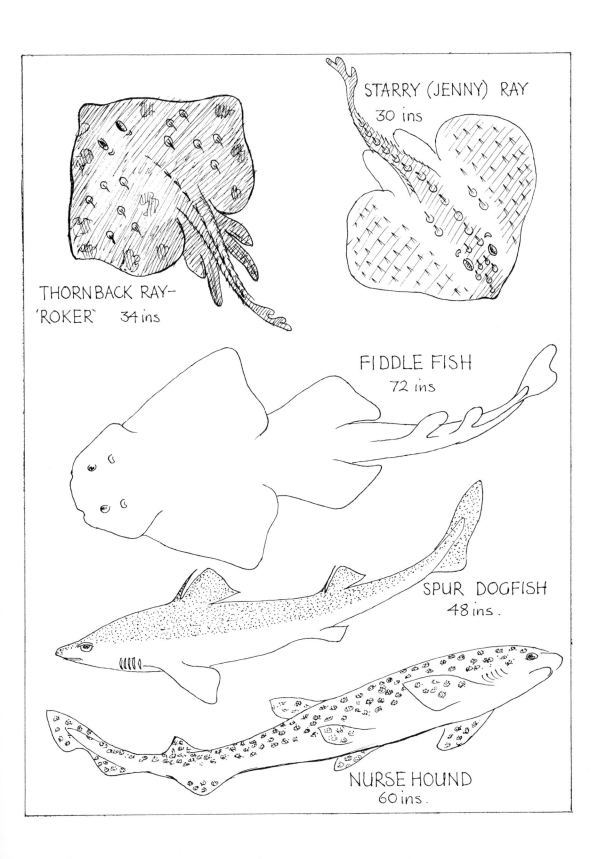

STARRY (JENNY) RAY
30 ins

THORNBACK RAY-
'ROKER' 34 ins

FIDDLE FISH
72 ins

SPUR DOGFISH
48 ins.

NURSE HOUND
60 ins.

Roads	the sea approaches to a port
Roller-cleat	cleat with a roller built into it
Roller-gang	space in a smack's rail, with a roller at the bottom, through which the warp ran freely
Rolls	roes
Roker/rooker	general term for rays
Rope-room	space below decks where the trawl warp lay coiled
Ross	hard sandy deposits, resembling coral in shape, that are built up by various marine worms
Rough stuff	any trawl-fish that isn't classed as prime
Rumpy	coral
Sausage nets	lengths of net on a hake trawl's headline, containing glass floats
Scruff	shell debris etc. on the sea bed
Scuffle	a bit of chop and turbulence on the surface of the sea
Scuppers/scuttles	holes in a ship's gunwale which allow water to drain away off the deck
Sheaves	grooved wheels in blocks that allow ropes to run freely
Sheer	(1) clear; (2) veer to port or starboard; (3) upward sweep of a boat's decks towards the bows
Sheet	rope by which the set of a sail is controlled
Shiftenins	changes of underclothes
Ship's husband	man in charge of crewing arrangements and a boat's well-being
Shoe (kettle)	wedge-shaped metal container used for boiling water and making tea in an upright boiler
Shoot	to cast fishing nets
Sinkers	heavy dumplings
Skeleton ropes	ground-ropes that have no chain on them
Slabs	plaice that have recently spawned
Slink	mature cod that has very little flesh on it
Slip	small sole
Sole tub	large wooden tub that soles were washed in on board ship
Sparky	wireless operator
Spell	change over watches on board ship
Spinnaker	the largest of all the head sails
Sprags	half-grown cod
Standard boats	steam drifters and trawlers built during the First World War to Admiralty specification
Stay	rope or wire support to a mast
Stem	(1) the bow post of a boat; (2) to head the tide
Stockie bait	money paid to trawlermen out of the sale of small fish, gurnards etc.
Stopper	(1) plaited rope that held the warp to the towing post on a

	smack; (2) wire strop that linked an otter door to the net; (3) heavy chain on which a trawl door hung in the gallus
Store	building where all manner of fishing gear was kept and maintained
Swabs	brooms or mops for cleaning decks and dressing sails
Swimming fish	members of the cod/hake family
Tanner organs	scollops
Tanning	dressing sails
Third hand	man below mate on a smack or trawler (a Lowestoft term)
Thwarts	a dinghy's seats
Ticket	skipper's or mate's certificate of competence
Tide about	work the tides in order to get in as many tows as possible
Titlers	chains on a trawl that cause sanding-up
Tow-dinger/tow-fores'l	a large triangular fore sail
Towing block	device that held trawl warps together while fishing was in progress
Trimmer	crewman on board a trawler who prepared coal for the furnace and tended the lights
Truck	(1) the top of a mast; (2) the bottom hoop on a smack's sail
Trunk	wooden case for packing trawl-fish in
Tusk	torsk
Watchman	person who kept an eye on fishing boats as they lay in harbour overnight
Wear away	let go trawl gear
Weather tide	a tide running against the wind
Weighings	quantities of fish totalling one cwt
Wells	compartments in a fish-hold
Wigging	listening in on the wireless. An abbreviation of ear-wigging
Will-ducks	guillemots
Wings	(1) spaces below deck between the hold and the sides of the boat; (2) parts of a trawl net nearest the doors or the beam
Wires	warps
Wrapper	silk neckerchief

Smacks 'shoe' kettle
-evolved for heating on capstan boiler furnace.
approx 18 ins.

Select Bibliography

The Sea Fisherman, or Sea Pilotage by J. Wilcock (Stephen Barbet, 1865)

Nor'ad of the Dogger by E.J. Mather (James Nisbet & Co., 1887)

The Victoria County History of Suffolk (Archibald Constable & Co. Ltd., 1907)

North Sea Fishers and Fighters by Walter Wood
(Kegan Paul, Trench, Trubner & Co. Ltd., 1911)

An Account of the Fishing Gear of England and Wales by F.M. Davis (H.M.S.O., 1923)

The Distinguishing Features of Fish by C.N.H. (The Worshipful Company of Fishmongers, 1949)

Alde Estuary by W.G. Arnott (Norman Adlard & Co. Ltd., 1952)

Sailing Trawlers by Edgar J. March (Percival Marshall & Co. Ltd., 1953. Reprinted by David & Charles 1970)

Trawlermen's Handbook compiled by Lieut.-Cdr. R.C. Oliver
(Fishing News Books Ltd., 1965)

Modern Deep Sea Trawling Gear by John Garner (Fishing News Books Ltd., 1967)

The Fishes of the British Isles & N.W. Europe by Alwyne Wheeler (Macmillan, 1969)

The Deep Sea Fishermen by Alan Villiers (Hodder & Stoughton, 1970)

The Roaring Boys of Suffolk by P. Cherry & T. Westgate (Brett Valley Publications, 1970)

Ports of the Eastern Counties by Wilfrid J. Wren (Terence Dalton Ltd., 1976)

Business in Great Waters by John Dyson (Angus & Robertson, 1977)

Sailing Fishermen in Old Photographs by Colin Elliott (Tops'l Books, 1978)

Steam Fishermen in Old Photographs by Colin Elliott (Tops'l Books, 1979)

The Cod Bangers by Hervey Benham (Essex Counties Newspapers Ltd., 1979)

The Driftermen by David Butcher (Tops'l Books, 1979)

Olsen's Fisherman's Nautical Almanac (E.T.W. Dennis & Sons Ltd. – published annually)

Toilers of the Deep (Quarterly Journal of The Royal National Mission to Deep Sea Fishermen)

Fishing News (Weekly publication by Arthur J. Heighway Publications Ltd., Fleet Street, London)

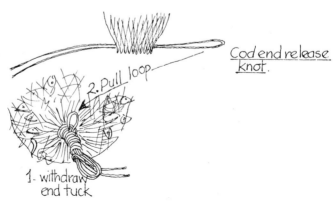

Cod end release knot.

2. Pull loop

1. withdraw end tuck

Back cover: Steam drifter-trawler pitching into the North Sea. As her fore galluses are not rigged she is probably going herring catching.